ollins

Maths Frameworking

3rd edition

Peter Derych, Kevin Evans,
Keith Gordon, Michael Kent,
Trevor Senior, Brian Speed

Published by Collins
An imprint of HarperCollins*Publishers*
The News Building
1 London Bridge Street
London
SE1 9GF

Browse the complete Collins catalogue at
www.collins.co.uk

ISBN-13 978-0-00-753765-5

British Library Cataloguing in Publication Data
A Catalogue record for this publication is available from the British Library.

Written by Peter Derych, Kevin Evans, Keith Gordon, Michael Kent, Trevor Senior, Brian Speed
Commissioned by Katie Sergeant
Project managed by Elektra Media Ltd
Development edited and copy-edited by Susan Gardner
Edited by Helen Marsden
Proofread by Joanna Shock
Illustrations by Ann Paganuzzi
Typeset by Jouve India Private Limited
Page and cover design by Angela English

With thanks to Chris Pearce
Printed and bound by L.E.G.O S.p.A. Italy

Acknowledgements
The publishers wish to thank the following for permission to reproduce photographs. Every effort has been made to trace copyright holders and to obtain their permission for the use of copyright materials. The publishers will gladly receive any information enabling them to rectify any error or omission at the first opportunity.

Cover Hupeng/Dreamstime

Contents

How to use this book

Welcome to *Maths Frameworking 3rd edition Homework Book 3*

Maths Frameworking Homework Book 3 accompanies Pupil Books 3.1, 3.2 and 3.3 and has hundreds of practice questions at different levels to help you consolidate what you have learned in class. Key features enable easy navigation through the book: indicators help you find the right questions for your level and question type icons help you find and practise key skills.

These are the key features:

Numbered topics match the Pupil Books so you can find the right sections easily.

The **level of difficulty** of the questions corresponds to the three different year 2 Pupil Books:

3.1 3.2 3.3

Practise your **problem solving, mathematical reasoning** and **financial skills** with highlighted questions.

(PS) (MR) (FS)

Challenge yourself with extended **Brainteaser** activities.

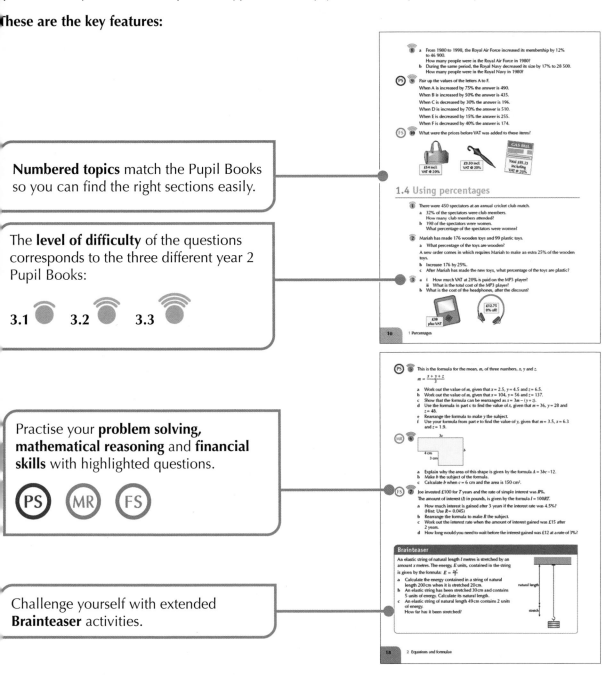

1 Percentages

1.1 Simple interest

  1 Work out these percentages. Do not use a calculator.

    a 60% of £800   **b** 5% of £250   **c** 20% of £150

    d 19% of £700   **e** 75% of £600   **f** 120% of £90

  2 Write down 25% of each quantity. Do not use a calculator.

    a £28   **b** 44 g   **c** 60 cm

    d 12 m   **e** $36   **f** 24 hours

  3 Work out these percentages of £80. Do not use a calculator.

    a 50%   **b** 10%   **c** 25%

    d 20%   **e** 75%   **f** 70%

  4 Use a calculator to work out these percentages.

    a 43% of £115   **b** 17% of £34   **c** 78% of £930

    d 2.6% of £85   **e** 11.2% of £560   **f** 0.4% of £45

  5 Cassylda takes out a loan of £600.

She pays simple interest of 5% per month for eight months.

    a Work out the amount of interest Cassylda pays each month.

    b Work out the total amount of interest Cassylda pays.

  6 Taiwo takes out a loan of £3200.

She pays simple interest of 2.8% per month for 10 months.

    a Work out the amount of interest Taiwo pays each month.

    b Work out the amount of interest Taiwo has paid after six months.

    c Work out the total amount of interest Taiwo has paid after 10 months.

> **LOANS**
> Only
> **2.8%**
> per month

  7 Laurel takes out a loan of £5250.

She pays simple interest of 3.7% per month for three years.

    a Work out how much interest she pays in the first year.

    b Work out how much interest she pays altogether.

8 Valerie takes out a loan of £2800. Shannon takes out a loan of £4600. Ruby takes out a loan of £5400.

All three pay 7% simple interest each year for six years.

a Calculate how much interest Valerie has to pay altogether.

b Calculate how much more interest Ruby has to pay altogether than Shannon.

  9 Rebecca takes out a loan of £630 and pays 2.1% simple interest every week for 12 weeks.

Ntuse takes out a loan of £720 and pays 1.7% simple interest every week for 14 weeks.

Kezia takes out a loan of £790 and pays 1.4% simple interest every week for 15 weeks.

Work out who pays the most interest.

  10 Angelie pays £16.08 monthly interest on a loan of £670.

Work out the rate of interest.

  11 Peggy pays £9.35 weekly interest on a loan of £550.

Ciara pays £8.93 weekly interest on a loan of £470.

Work out who pays the higher rate of interest and by how much.

  12 Aaliyah takes out a loan of £3600 to buy a new car.

She pays £691.20 simple interest monthly for two years.

a Work out the rate of interest.

b Work out her total interest payment as a percentage of the original loan.

Brainteaser

Elspeth took out a loan.

She paid 2.2% simple interest monthly for eight months, then 3.5% simple interest monthly for the next eight months.

The total amount of interest she paid was £1710.

How much was the loan?

1.2 Percentage increases and decreases

  1 a What is 10% of 40?

b Increase £40 by 10%.

c Decrease £40 by 10%.

 2 A book sells 2600 copies in a week. The book wins an award and sales increase by 50% the week after. How many copies does the book sell that week?

3 Elilini has £350 in a savings account.

How much will she have if her savings:

a increase by 20% **b** decrease by 20% **c** increase by 12%

d decrease by 12% **e** increase by 65% **f** decrease by 65%?

  4 VAT of 20% is added to each of these items. Work out the price after VAT is added.

a a coat (£140) **b** a toy helicopter (£26)

c a set of saucepans (£58) **d** a tent (£340)

  5 All items in a sale are reduced by 30%. Find the sale price for each item.

a a wardrobe (£470) **b** a bath (£350)

c a sofa (£1400) **d** a bed (£820)

  6 The price of a ticket for a musical was £85.

Work out the new price after:

a an increase of 16% **b** a decrease of 7%

c an increase of 25% **d** a decrease of 15%.

 7 Work out how much each person now earns.

a Magisha earnt £460 per week and then received a 4% pay rise.

b Fiona earnt £385 per week and then received a 2% pay cut.

c Gloria earnt £514 per week and then received a 9.1% pay cut.

d Anna earnt £193.28 per week and then received a 22% pay rise.

8 a When Kelly was calculating a percentage change she multiplied the original value by 1.4.

State whether this was an increase or a decrease and what the percentage change was.

b When Elaine was calculating a percentage change she multiplied the original value by 0.56.

State whether this was an increase or a decrease and what the percentage change was.

c When Simrath was calculating a percentage change she multiplied the original value by 0.13.

State whether this was an increase or a decrease and what the percentage change was.

d When Suzie was calculating a percentage change she multiplied the original value by 2.8.

State whether this was an increase or a decrease and what the percentage change was.

  9 The value of a vintage car is £28 000.

Work out the new value if it increases by:

a 10% **b** 80% **c** 150% **d** 250%.

10 In one year, a eucalyptus tree grew from 2.7 m tall to 4.3 m tall. What was the percentage increase in height?

 11 Frances filled her petrol tank with 60 litres of petrol. After a journey, there were 18 litres left. What was the percentage decrease in fuel?

 12 This table shows how the populations of some towns and villages changed between 1980 and 2000. Copy and complete the table.

Place	Percentage change	Population in 1980	Population in 2000
Smallville	30% decrease		1300
Lansbury	8% increase	46 000	
Gravelton	2% decrease		19 800
Smithchurch	48% increase		144 000
Deanton		28 000	33 320
Tanwich		680	646

Brainteaser

Sometimes when somebody wants to buy an expensive item, they can pay using 'hire purchase' which means that they pay a percentage each month and do not own the item until after the last payment is made.
Romain and Anais wish to buy computers using hire purchase.

Romain's computer costs £1200 and his first payment was on 1st March 2014.

Anais' computer cost £1500 and her first payment was on 1st July 2014.

Romain pays 2.5% per month.

Anais pays 4% per month.

What will be the final dates of each of Romain and Anais' last payments?

In which month will they have the same amount left to pay?

1.3 Calculating the original value

 1 **a** After a 50% increase, a price is now £30. Work out the original price.
b After a 50% decrease, a price is now £30. Work out the original price.

 2 Write down the multiplier for each percentage increase.

a 22%	**b** 48%	**c** 6%
d 8.5%	**e** 120%	**f** 333%

3 Write down the multiplier for each percentage decrease.

    a 22%   **b** 48%   **c** 6%   **d** 8.5%

4   **a** A price is £105 after a 25% increase. Work out the original price.
      b A price is £243 after a 25% decrease. Work out the original price.
      c A price is £234 after a 4% increase. Work out the original price.
      d A price is £648 after a 4% decrease. Work out the original price.

5 The average water consumption of a household is 165 litres per day.
This is an increase of 10% on last year.
What was the consumption last year?

6 There are 56 spectators at a hockey match.
This is 20% fewer than at the last match.
How many spectators were at the last match?

7 A motorway has been extended to 224 miles.
This is a 28% increase from last year.
How long was the motorway before?

8 Emily now weighs 70.3 kg.
This is a 5% decrease from six months ago.
What did Emily weigh six months ago?

9   **a** From 1980 to 1998, the Royal Air Force increased its membership by 12% to 46 900.
      How many people were in the Royal Air Force in 1980?
     b During the same period, the Royal Navy decreased its size by 17% to 28 500.
      How many people were in the Royal Navy in 1980?

10 Pair up the values of the letters A to F.
When A is increased by 75% the answer is 490.
When B is increased by 50% the answer is 435.
When C is decreased by 30% the answer is 196.
When D is increased by 70% the answer is 510.
When E is decreased by 15% the answer is 255.
When F is decreased by 40% the answer is 174.

11 What were the prices before VAT was added to these items?

£54 incl.
VAT @ 20%

£9.50 incl.
VAT @ 20%

GAS BILL

Total £85.23
including
VAT @ 20%

1.4 Using percentages

1 There were 450 spectators at an annual cricket club match.

    a 32% of the spectators were club members.
       How many club members attended?

    b 198 of the spectators were women.
       What percentage of the spectators were women?

2 Mariah has made 176 wooden toys and 99 plastic toys.

    a What percentage of the toys are wooden?

    A new order comes in which requires Mariah to make an extra 25% of the wooden toys.

    b Increase 176 by 25%.

    c After Mariah has made the new toys, what percentage of the toys are plastic?

(FS) 3   **a** **i** How much VAT at 20% is paid on the MP3 player?
         ii What is the total cost of the MP3 player?

    b What is the cost of the headphones, after the discount?

£12.75
8% off!

£38
plus VAT

(FS) 4 Abigail runs a website that sells theatre tickets.

    She calculates that sales for child and adult tickets in March were in the ratio 3 : 5.

    a What percentage of the ticket sales were adult tickets?

    b Work out the number of child tickets sold as a percentage of adult tickets sold.

    In March, 261 child tickets were sold.

    c How many adult tickets were sold?

    Child tickets cost £9 each and adult tickets cost £14 each.

    d Find the total ticket sales on Abigail's website in March.

5 The table shows the numbers of members of a club by gender and age.

	Male	Female
Under 25	27	45
25 or over	72	36

    a What percentage of the members are females under 25?

    b What percentage of the members are male?

    c What percentage of the members are 25 or over?

6 The diameter of the planet Mercury is 38% of the diameter of Earth.

The diameter of the Earth is 12 756 km.

a Round 38% to 1 significant figure.
b Round 12 756 km to 1 significant figure.
c Use your approximations to estimate the diameter of Mercury.

 7 The diagram shows the heights of two trees a year ago.

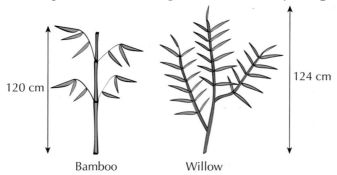

Bamboo   Willow

During the last year, the bamboo tree grew 10% and the willow tree grew 5%.

Which is the taller tree now?

  8 A garden centre is holding a sale where all items have been reduced by the same percentage.

Sarah buys a wheelbarrow for £28.50 which has been reduced from £37.50.

a Find the percentage reduction.

Simi buys a lawnmower which cost £67.50 before the sale.

b Find the sale price of the lawnmower.

Izzy buys a bag of compost for £17.10.

c Find the price of the bag of compost before the sale.

  9 216 of the cushions made in a factory each day are red.

This is 45% of the total number of cushions.

a How many cushions are made in total in one day?

One third of the cushions are blue and the rest are yellow.

b How many yellow cushions are made in five days?

The factory makes £6 profit for each red cushion, £8 profit for each blue cushion and £10 profit for each yellow cushion.

c What percentage of the factory's profits comes from red cushions?

Brainteaser

Steph deposits £3200 into a bank account.

The account pays her interest each year of 2.6%.

a How much money will there be in the account after one year?
b How much money will there be in the account after three years?
c How much interest will have been added after five years?

2 Equations and formulae

2.1 Expansion

1. Simplify these expressions.

   a $3 \times 4p$ **b** $5 \times 6q$ **c** $8r \times -2$ **d** $-7 \times -5t$

2. Expand the expressions.

   a $5(d + 2)$ **b** $7(2p - 1)$ **c** $4(3 + m)$ **d** $10(5 - 3i)$

3. Expand these expressions.

   a $-(3a - 2)$ **b** $-4(-2h + 6)$ **c** $-2(2w - 3)$ **d** $-6(-5 - 6x)$

(PS) 4. **a** The sides of a square all measure $3s + 5$. Write an expression for its perimeter. If you use brackets, expand the expression and simplify.
   b If $s = 7\,\text{cm}$, what is the perimeter of the square in metres?

5. Expand and simplify the expressions.

   a $4(x + 5) - 2x$ **b** $5(2y + 1) + 2y - 3$ **c** $-3(t + 4) + 7t$
   d $10 - 5(x - 1)$ **e** $9g - 2(g + 3)$ **f** $p - 4(2 - p)$

6. Expand both brackets and simplify the expressions.

   a $4(x + 3) + 2(x - 2)$ **b** $2(i - 1) + 3(2i + 1)$ **c** $5(2n - 3) + 3(4n + 1)$
   d $4(r + 2) - 2(r - 3)$ **e** $3(2k + 3) - 2(k + 4)$ **f** $4(3j - 4) - 5(2j - 3)$
   g $2(3 - 2f) + 5(4 + 3f)$ **h** $3(6 + u) - 4(3 - 5u)$

(PS) 7. Make a sketch of this flag.

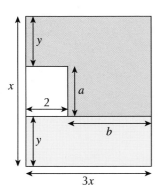

   a Find an expression for **i** a and **ii** b and mark them on your diagram.
   b Find expressions for:

   i the yellow area   **ii** the white area   **iii** the blue area.

Brainteaser

A family orders fish and chips four times from Jack's Chip Shop, which charges £2.80 for fish and £1.20 for chips.

a How much does the meal cost?

b Write this order as a formula, using f for fish, c for chips, and a pair of brackets.

c What percentage of the cost goes on the fish?

d Harry's Fish Bar charges £16.80 for the same meal. If chips are £1.25, how much do they charge for fish?

e What is the ratio of the prices between the two shops?

f How much more expensive is Harry's Fish Bar than Jack's Chip Shop as a percentage?

2.2 Factorisation

1 Work out the highest common factor for each pair of numbers.

    a 15 and 20   **b** 12 and 30   **c** 8 and 24   **d** 18 and 36

2 Factorise the expressions.

    a $5m + 20$ **b** $3x - 21$ **c** $35 + 7k$ **d** $11 - 55h$

3 Factorise the expressions.

    a $8m - 12n$ **b** $6p + 9q$ **c** $27a - 18b$ **d** $-24f - 16g$

4 Simplify each expression by collecting like terms. Then factorise it.

    a $2a + 5 + 3a + 10$ **b** $9x - 8 + x - 2$ **c** $12y - 3 + 18y + 9$

    d $15x + 2 - 11x - 10$ **e** $p + 23 + 4p - 8$ **f** $6 - 3y - 12 - 3y$

5 The perimeter of an equilateral triangle measures $9x + 15$. What is the length of each side?

PS **6** **a** If the units digit of a number is a, give an expression for:

        i the tens column   **ii** the hundreds column.

    b Find an expression for any two-digit number where the tens digit is twice the units digit, e.g. 63.

    c Find more two-digit numbers with this property.

    d 63 is divisible by 21. Simplify your expression to show that numbers where the tens digit is twice the units digit are always divisible by 21.

7 **a** For each shape, write an expression for the perimeter.

b Then simplify it as much as possible, and factorise it if you can.

i   **ii**   **iii**

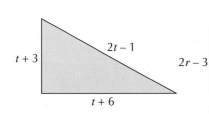

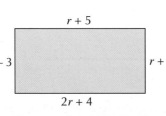

$t + 3$ $2t - 1$ $t + 6$

$r + 5$ $2r - 3$ $r + 3$ $2r + 4$

$p + 5$ $p + 5$ $p + 8$ $p + 8$ $p + 4$

Brainteaser

In the block of flats shown, each small window is x metres wide, and the large window is $2x$ metres wide, with 1 m on each side of each window. The windows are all y metres high with 2 m above and below each window.

a Write factorised expressions for

  i the overall width of the building

  ii the height of the building.

b What are the dimensions of the block when $x = 1.3$ m, and $y = 1.8$ m?

c If the block measures 20 m wide by 9.6 m high, what are the values of x and y?

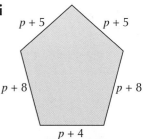

Not drawn to scale

2.3 Equations with brackets

1 Solve these equations.

  a $5w + 20 = 35$ **b** $12x - 18 = 0$ **c** $30 + 8y = 14$ **d** $14 - 21z = -7$

2 Solve these equations. Show your method.

  a $2(y - 8) = 30$ **b** $3(6 + k) = 18$ **c** $4(y + 3) = 36$ **d** $3(8 - y) = 15$

  e $4(3 + t) = 52$ **f** $5(6 - p) = 45$ **g** $40 = 5(f - 17)$ **h** $24 = 4(w - 3)$

3 Solve these equations. Give your answers in fraction form.

  a $3(x - 7) = 5$ **b** $4(y + 1) = 13$ **c** $6(t - 1) = 27$ **d** $9(c + 3) = 28$

4 **a** Simplify the expression $5(x + 2) + x - 3$.

  b Now solve the equation $5(x + 2) + x - 3 = 37$.

5 **a** Simplify the expression $7(x - 1) + 6(x - 2)$.

  b Now solve the equation $7(x - 1) + 6(x - 2) = 30$.

MR **PS** **6** **a** Explain why a regular hexagon whose perimeter is $12h + 6$ must have sides all equal to $2h + 1$.

  b What must the sides and perimeter of the hexagon be if $h = 3.5$ cm?

  c What is the value of h if the hexagon's perimeter is 30 cm?

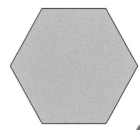

7 **a** Show that the expression $5(x - 3) - 4(x - 2)$ simplifies to $x - 7$.
   b Use the result in part **a** to find a quick solution to the equation
   $5(x - 3) - 4(x - 2) = 10$
   c Solve the equation $5(x - 3) - 4(x - 2) = -7$

8 Solve these equations and show your working.

   a $3(m - 1) + 4(m + 2) = 52$ **b** $7(d + 2) - 2(2d + 1) = 21$
   c $2(6t + 1) - 4(t - 1) = 0$ **d** $4(2 - b) + 3(3 + 2b) = 33$

9 Solve these equations and show your working.

   a $7(v + 3) = 5(v + 7)$ **b** $2(2d + 4) = 4(2d - 7)$
   c $3(x - 2) = -2(x + 19)$ **d** $2(5 - 2k) = 3(1 + 2k)$

2.4 Equations involving fractions

In questions **1** to **4**, solve the equations. Show your working.

1 **a** $\frac{x}{2} = 8$ **b** $\frac{x}{3} = 7$ **c** $\frac{x}{10} = -5$ **d** $\frac{m}{0.4} = 0.3$

2 **a** $\frac{x}{2} - 3 = 4$ **b** $\frac{x}{4} + 2 = 12$ **c** $\frac{x}{3} - 6 = -5$ **d** $\frac{v}{3} + 7 = -10$

3 **a** $\frac{3d}{5} = 6$ **b** $\frac{4u}{3} = -5$ **c** $-\frac{3m}{5} = 9$ **d** $\frac{2t}{-3} = -5$

4 **a** $\frac{x + 2}{3} = 2$ **b** $\frac{y - 6}{5} = 1$ **c** $\frac{2m + 3}{3} = 5$ **d** $\frac{5t - 3}{4} = -2$

   e $\frac{1}{4}(x + 6) = 5$ **f** $\frac{1}{3}(x - 8) = 2$ **g** $\frac{1}{6}(x + 12) = -2$ **h** $\frac{1}{8}(x - 4) = -3$

5 The width of this rectangle is 5 cm and the area is $3n + 4$ cm².

5 cm $3n + 4$ cm²

   a Explain why the length of the rectangle is $\frac{3n+4}{5}$ cm.
   b You are told that the length of the rectangle is 13 cm.
      Write down an equation and solve it to work out the value of n.

6 Solve these equations.

   a $\frac{2}{3}(x + 2) = 8$ **b** $\frac{3}{4}(t - 1) = 2$ **c** $\frac{2}{5}(z + 5) = 7$ **d** $\frac{3}{5}(r - 2) = 4$

   e $\frac{2}{5}(a + 3) = 5$ **f** $\frac{3}{8}(t + 10) = 6$ **g** $\frac{3}{8}(z - 2) = 4$ **h** $\frac{3}{8}(m - 11) = 2$

7 Solve these equations. Leave your answers as simplified fractions.

a $\dfrac{2(x+5)}{3} = 16$ **b** $\dfrac{3(t-2)}{4} = 4$ **c** $\dfrac{3(2+x)}{10} = 4$ **d** $\dfrac{5(c-10)}{9} = 2$

e $\dfrac{6(r-8)}{5} = 2$ **f** $\dfrac{2(3+t)}{5} = 4$ **g** $\dfrac{9(d-2)}{10} = 2$ **h** $\dfrac{3(e-5)}{4} = 4$

Brainteaser

Four numbers are x, $x+3$, $x-5$, and $2x$.

a Given that $x = 8$, work out the value of the four numbers.
b Show that the mean of the four numbers is $\frac{5x-2}{4}$.
c Suppose the mean of the four numbers is 12.
  Write down an equation and solve it to work out the value of x.
d Work out the four numbers for the value of x you found in part **c**.
e What is the range of the four numbers?

2.5 Rearranging formulae

1 **a** Make m the subject of the formula $L = m + n$.
  b Make h the subject of the formula $A = bh$.
  c Make s the subject of the formula $v = \frac{s}{t}$.
  d Make t the subject of the formula $v = \frac{s}{t}$.
  e Make R the subject of the formula $D = x - R$.

2 This is a formula that you will see in science: $v = u + at$.

  a Work out the value of v, given that $u = 15$, $a = 9.8$ and $t = 5$.
  b Rearrange the formula to make t the subject.
  c Work out the value of t when $u = 0$, $v = 4.5$, and $a = 12$.
  d Work out the value of t when $u = 10$, $v = 8$, and $a = 7$.

3 The time, T minutes, to roast a joint of meat weighing W grams is given by the
  formula: $T = 20 + \dfrac{W}{25}$.

  a Calculate the time needed to roast a joint of meat weighing 800 g.
  b Calculate the time needed to roast a joint of meat weighing 1.5 kg.
  c Make W the subject of the formula.
  d Calculate the weight of a joint that needs 1 hour 30 minutes of roasting time.

4 The total number of legs, L, in a classroom is given by the formula:

  $L = 4c + 2p$ where c is the number of chairs and tables and p is the number of people.

  a Make p the subject of the formula.
  b Use your formula to calculate the number of people in a classroom containing
    16 chairs and 6 tables given that there are 112 legs altogether.

(PS) **5** This is the formula for the mean, m, of three numbers, x, y and z.

$$m = \frac{x + y + z}{3}$$

  a Work out the value of m, given that $x = 2.5$, $y = 4.5$ and $z = 6.5$.
  b Work out the value of m, given that $x = 104$, $y = 56$ and $z = 137$.
  c Show that the formula can be rearranged as $x = 3m - (y + z)$.
  d Use the formula in part **c** to find the value of x, given that $m = 36$, $y = 28$ and $z = 48$.
  e Rearrange the formula to make y the subject.
  f Use your formula from part **e** to find the value of y, given that $m = 3.5$, $x = 6.3$ and $z = 1.9$.

(MR) **6**

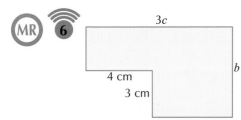

  a Explain why the area of this shape is given by the formula $A = 3bc - 12$.
  b Make b the subject of the formula.
  c Calculate b when $c = 6$ cm and the area is 150 cm².

(FS) **7** Joe invested £100 for T years and the rate of simple interest was R%.

The amount of interest (I) in pounds, is given by the formula $I = 100RT$.

  a How much interest is gained after 3 years if the interest rate was 4.5%? (Hint: Use $R = 0.045$)
  b Rearrange the formula to make R the subject.
  c Work out the interest rate when the amount of interest gained was £15 after 2 years.
  d How long would you need to wait before the interest gained was £12 at a rate of 3%

Brainteaser

An elastic string of natural length l metres is stretched by an amount x metres. The energy, E units, contained in the string is given by the formula: $E = \frac{2x^2}{l}$

  a Calculate the energy contained in a string of natural length 200 cm when it is stretched 20 cm.
  b An elastic string has been stretched 30 cm and contains 5 units of energy. Calculate its natural length.
  c An elastic string of natural length 49 cm contains 2 units of energy. How far has it been stretched?

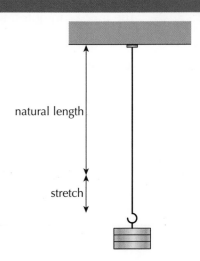

natural length

stretch

3 Polygons

3.1 Angles in polygons

1. What is the name of each of these polygons?
   - **a** eight vertices
   - **b** seven sides
   - **c** five sides
   - **d** ten vertices

2. How many vertices does each polygon have?
   - **a** hexagon
   - **b** dodecagon
   - **c** nonagon
   - **d** parallelogram

3. An icosagon polygon has 20 sides.
   - **a** How many equal triangles can an icosagon be divided into?
   - **b** What is the sum of its interior angles?

4. The sum of the interior angles of a polygon is 2520°.
   - **a** How many equal triangles can the polygon be divided into?
   - **b** How many sides does the polygon have?

5. The interior angles of a hexagon are 100°, 150°, 75°, 90°, 145° and x. Find x.

6. Calculate the size of the unknown interior angle.

   i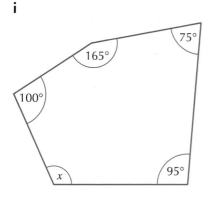

   ii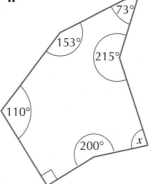

   iii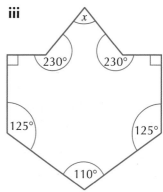

7 Calculate the size of the unknown exterior angle.

i

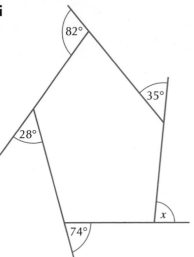

ii

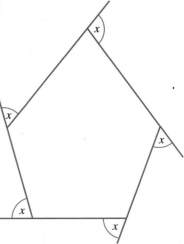

8 Find interior angle x.

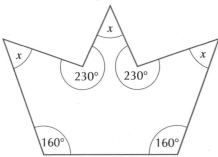

MR **9** **a** Calculate the marked angles. Explain how you found each angle.

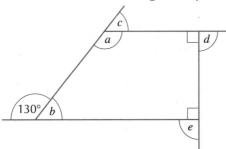

b Check that the exterior angles total 360°.

10 The five interior angles of a pentagon are $6x - 2$, $7x + 1$, $4x + 5$, $2x + 80$ and $8x - 3$.
Calculate the size of each interior angle of the pentagon.

11 A polygon has 50 sides.
All but one of its angles is 170°.
What is the size of the other angle?

12 How many sides does each of these polygons have?
  a Sum of the interior angles = 6660°
  b Sum of the interior angles = 9540°
  c Sum of the interior angles = 12 780°

3 Polygons

3.2 Angles in regular polygons

1.
   a. What is the sum of the angles in a triangle?
   b. What is the special name given to a regular triangle?
   c. What is the size of each interior angle in a regular triangle?

2.
   a. What is the sum of the angles in a quadrilateral?
   b. What is the special name given to a regular quadrilateral?
   c. What is the size of each interior angle in a regular quadrilateral?

3. A regular nonagon has nine sides.
   a. How many equal triangles can this polygon be divided into?
   b. What is the sum of its interior angles?
   c. What is the size of each interior angle?

4. A regular polygon has 12 sides.
   a. How many equal triangles can the polygon be divided into?
   b. What is the sum of its interior angles?
   c. What is the size of each interior angle?

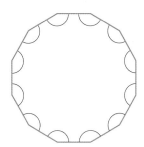

5. Find the sum of the exterior angles of a regular polygon with 13 sides.

 6. Copy and complete this table for regular polygons. Include reasons for your answers in the table.

Number of sides	Sum of exterior angles	Size of each exterior angle	Size of each interior angle
15			
16			
18			
20			
30			

7. Find the number of sides of a regular polygon with an interior angle of
   a. 170°   b. 179°   c. 179.9°   d. 179.999°

8.
   a. Calculate the marked angles for this regular hexagon.
   b. Prove that ABDE is a rectangle.

   (Hint: Show that its angles are all 90°.)

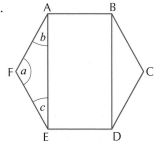

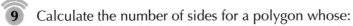

9 Calculate the number of sides for a polygon whose:

    a exterior angle is 4°

    b interior angle is 171°

    c interior angles add up to 1980°

10 Calculate the marked angles for this regular octagon.

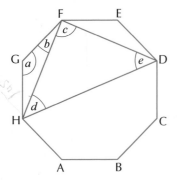

11 This diagram shows a regular pentagon and a regular hexagon joined together. Calculate the marked angle.

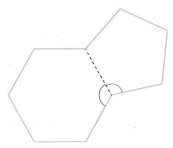

12 Calculate the marked angles for this regular heptagon. Give your answers correct to one decimal place.

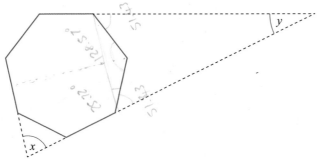

3.3 Regular polygons and tessellations

  1 a Copy each triangle on to squared paper. Does each triangle tessellate?

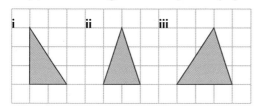

    b Draw a scalene triangle (three unequal sides) of your own on squared paper. Does it tessellate?

    c Does a triangle always tessellate? Explain your answer.

  2 a Copy each quadrilateral on to squared paper. Does each quadrilateral tessellate?

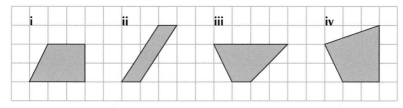

    b Draw a quadrilateral of your own on squared paper (not a special quadrilateral). Does it tessellate?

    c Does a quadrilateral always tessellate? Explain your answer.

 3 Draw six more copies of the kite on squared paper and show how it could tessellate.

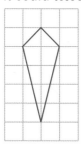

4 Draw six more copies of the arrowhead on squared paper and show how it could tessellate.

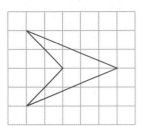

5 Show how a regular hexagon can tessellate.
You should draw at least 10 hexagons.

6 Draw a tessellation for this shape.

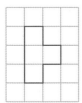

7 Draw a tessellation for this shape.

MR **8** **a** Copy and complete the table to state the interior and exterior angle for each regular shape.

Polygon	Interior angle	Exterior angle
triangle		
quadrilateral		
pentagon		
hexagon		
heptagon		
octagon		
nonagon		
decagon		

b For a regular polygon to tessellate, there need to be a whole number of polygons around a single point. Use the table to explain why only the regular triangle, quadrilateral and hexagon can tessellate.

4 Using data

4.1 Scatter graphs and correlation

1 Ten identical houses had their lofts converted. These scatter graphs show the number of workers, time taken and cost of equipment hire for each loft.

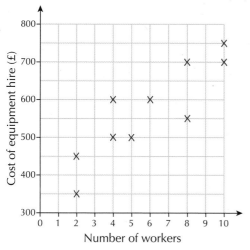

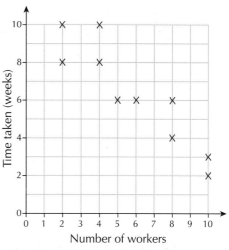

  a Describe the correlation between the number of workers and the time taken.

  b Describe the correlation between the number of workers and the cost of equipment hire.

  c **i** Describe the correlation between the time taken and the cost of equipment hire.

     ii Sketch a scatter graph to illustrate this correlation.

2 Ten pupils spent between 60 and 90 minutes writing a poem. Each poem was marked out of 100. What correlation would you expect between the time spent writing and the marks obtained?

3 The table shows the incomes of 10 people and the size of their car engines.

Income, £	10 000	14 000	17 000	27 000	30 000	36 000	37 000	42 000	45 000	50 000
Engine size, cc	1200	1800	1600	1700	2400	3200	1600	2700	3700	2800

  a Draw a scatter diagram using these axes.

    Horizontal axis (income): £10 000 to £50 000

    Vertical axis (engine size): 1000 cc to 4000 cc

  b Describe the correlation between income and engine size.

4 There are three ways of combining the variables x, y, z, in pairs (x with y, y with z and z with x). Write down what sort of correlation each pair will have.

  x: The number of people who travel to work by car each month

  y: The number of people who travel to work by public transport each month

  z: The amount of car exhaust pollution each month

  Illustrate each correlation by sketching a scatter graph.

5 This table shows the size and price of books sold by a bookshop in an hour.

Number of pages, n	60	60	120	140	140	180	240	280	320	400	480
Price of book, £P	1.50	1.00	2.50	3.00	3.20	5.00	4.50	6.00	6.00	7.50	10.00

a   Draw a scatter graph for the data. Use these scales.
  Number of pages: 2 cm to 100 pages
  Price of book: 1 cm to £1
b   Draw a line of best fit, by eye.
c   Use your line of best fit to estimate:
  i   the cost of a book containing 260 pages.
  ii the number of pages in a book costing £1.50.

 6 Fifteen young drivers were caught speeding through a village. Their ages and the speeds are shown in the table below.

Age	17	17	18	19	20	20	21	22	22	23	24	25	27	27	29
Speed (mph)	46	54	50	44	48	50	42	42	46	46	42	46	40	36	40

a   Draw a scatter graph for the data.
b   Use your graph to estimate:
  i   the age of a driver caught speeding at 45 mph.
  ii the speed of a driver aged 26.

Brainteaser

Hing Wai conducted an investigation into the amount of fiction and non-fiction adults watch on TV. He asked 10 of his teachers these questions.

'How many hours do you spend watching films each week?'

'How many hours do you spend watching documentaries each week?'

His report included this table of results and scatter diagram.

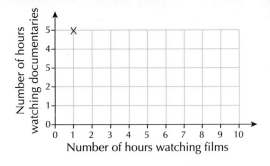

Teacher	Ms Witon	Mr Carter	Mr Singh	Mrs Tenby	Mrs Dean	Mr Chan	Ms Aldridge	Mr Jones	Ms Foster	Mr Phillips
Number of hours watching films	0	1	5	4	3	2	5	3	4	3
Number of hours watching documentaries	4	5	0	2	4	3	3	4	3	2

a   Copy and complete the scatter diagram.
b   Draw a line of best fit on your diagram.
c   Describe any correlation.
d   Use your line of best fit to estimate
  i   the number of hours spent watching films when $3\frac{1}{2}$ hours are spent watching documentaries.
  ii the number of hours spent watching documentaries when 9 hours are spent watching films.
e   What conclusion could Hing Wai make, based on this data?
f   Give a reason why this conclusion may not be true.
g   Write down another useful question that Hing Wai could have asked.

4.2 Time-series graphs

1 The normal depth of a pool is 1 metre. The graph shows the depth of water in an outdoor swimming pool at the end of each of the first 11 days of June.

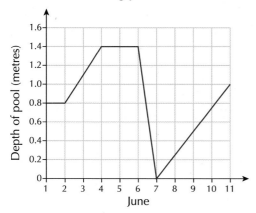

a On which day was the pool emptied?

b It rained on two consecutive days. When did it rain?

c **i** How long did it take to refill the pool to the normal depth?

  ii By how much did the depth increase per day?

c When was the rate of change in the depth of the pool the greatest?
Give a reason for your answer.

2 The graph below shows the rates of unemployment in the United Kingdom over the 20th century.

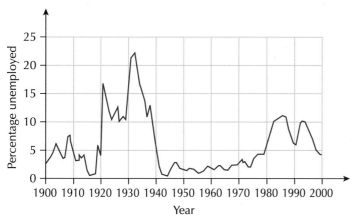

a In which decade was the rate of unemployment highest?

b Estimate the percentage of unemployed people in 1980.

c There were two periods when the rate of unemployment was close to zero.
Why do you think there was little unemployment during these periods?

d Pierre says, 'Unemployment was about the same at the beginning of the century as at the end'. Is he correct?

e The Second World War ended in 1945. For approximately how many years was the unemployment rate below 3% after the Second World War?

3 This diagram shows how the USA produced its energy during the 20th century.

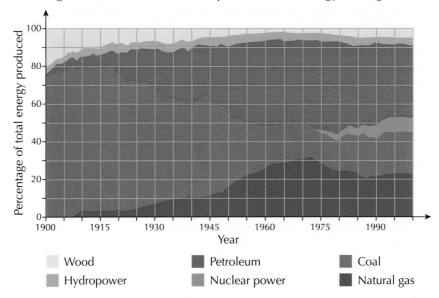

a What was the main source of energy at the:
   i beginning of the century   ii end of the century?
b Which source of energy provided roughly the same percentage of energy throughout the century?
c Estimate when petroleum and coal were used in equal amounts.
d What percentage of US energy was produced from natural gas in 1975?
e What percentage of US energy was produced from coal in 1945?
f Describe the trend of using wood as an energy source.
g Which energy source provided about twice as much energy as hydropower at the end of the century?
h Describe the trend in the use of natural gas.
i The total energy consumed in 1990 was 82 quadrillion units.
   i Find out what a quadrillion is.
   ii Estimate the energy produced from petroleum in 1990.

4 These graphs show the prices and sales of shares in two companies, James, James and Yates (JJY) and Standard Wire (SW), during a day of trading. The number of shares sold is called the 'volume'.

The prices shown are at the end of each 30-minute period. For example, the price of SW was $11.50 at 4:00 pm, and 20 000 shares were sold between 3:30 pm and 4:00 pm.

a Which share started trading at the highest price of the day?
b Which share has the greatest range of prices?
c What is the greatest price increase during a half-hour period? State the share and when it occurred.
d Bad news caused one of the shares to greatly reduce in price.
   i Which share was this?
   ii When did this happen?
   iii What was the fall in price?
e What was the difference in share prices at 12:30 pm?
f What do you notice about the volume during the day for both shares? Give a reason why this might have happened.
g During which hour was the fewest number of shares traded for each company?

h How much money was spent on JJY shares from 2:30 pm to 3:30 pm?

i During which half-hour period was most money spent altogether? How much was spent?

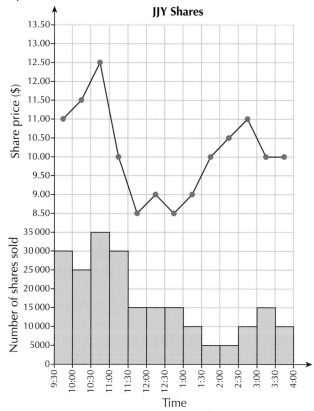

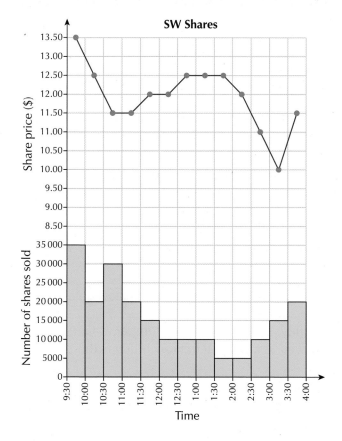

5 Look at the graph comparing the sale of printed books with downloaded e-books. The *x*-axis shows the year and alternate months.

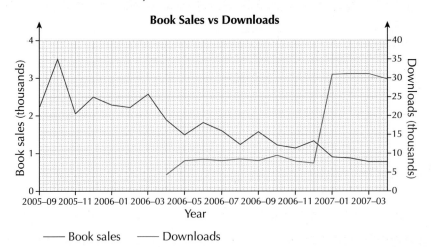

Book Sales vs Downloads

—— Book sales   —— Downloads

a In which month of which year did printed book sales reach their peak?
b When were the first e-books available, and how many were downloaded in the first month?
c How many months did it take before people started downloading in big numbers? Why do you think that was?
d What was the difference in sales of ordinary books and e-books in January 2007?
e Which month showed the biggest decrease in printed book sales? How do you know?
f Would it be fair to say that electronic downloads caused a massive reduction in the number of printed books sold?

4.3 Two-way tables

  1 A shop stocks the following range of jumpers.

	Brown	**Green**	**Blue**	**Pink**
Small	2	4	4	5
Medium	4	3	6	4
Large	4	4	5	3
Extra large	3	4	3	2

a What percentage of the jumpers stocked are extra large?
b What is the modal size?
c Although there are just four colours, only one makes up a quarter of the stock. Which one?
d What is the probability of choosing a blue jumper at random?
e The jumpers are priced at £9.50 each. How much is this stock worth in total?

Pupils at one school opt to take the following sports in their GCSE lessons.

Sport	Male	Female	Total
Basketball	12	15	
Swimming	15	18	
Tennis	16	14	
Total			

    a Copy and complete the totals for this table.
    b Which sport was preferred by a third of the pupils?
    c What percentage of girls chose swimming?
    d What is the probability that a boy chose basketball?
    e How many more chose swimming than tennis in percentage terms?

 This table shows details of all satellite launches during 2013 by launch site.

Site	Country	Launches	Successes	Failures	Partial failures
Kourou	France	7	7	0	0
Satish Dhawan	India	3	3	0	0
Tanegashima	Japan	2	2	0	0
Uchinoura	Japan	1	1	0	0
Jiuquan	China	7	7	0	0
Taiyuan	China	5	4	1	0
Xichang	China	3	3	0	0
Dombarovsky	Russia	2	2	0	0
Plesetsk	Russia	7	6	0	1
Naro	South Korea	1	1	0	0
Cape Canaveral	United States	10	10	0	0
Vandenberg	United States	4	4	0	0
Ocean Odyssey	International	1	0	1	0
Baikonur	Kazakhstan	23	22	1	0

    a How many launches took place in 2013?
    b What proportion were not successful? (Give your answer as a fraction and a percentage.)
    c Jag says, 'China launched more satellites than the USA.' Is he correct?
    d Which site was used for 12.5% of launches?

4 In a fell running race, the leading competitors gave their clubs and ages as follows:

Race Position	Club	Age
1	Pudsey Pacers	M45
2	Saddleworth Runners	M
3	Todmorden	M45
4	Saddleworth Runners	F
5	Unattached	M40
6	Pudsey Pacers	M55
7	Calder Valley	M50
8	Holmfirth Harriers	M60
9	Calder Valley	F40
10	Saddleworth Runners	M
11	Unattached	F45
12	Pudsey pacers	M45
13	Saddleworth Runners	F
14	Saddleworth Runners	M50
15	Holmfirth Harriers	F40

Key: M = male, F = female, age given are the youngest in a five year category, and where no age is given, runners are under 40.

a Put this data into a two-way table like this.

	Male	Female	totals
<40			
40–44			
45–49			
50–54			
55–59			
60+			
totals			

b What proportion of these runners were women?
c Team position is found by adding the first three runners from each club; lowest position wins. Which team won this race?
d An article in an athletics magazine claims that these results show fell running is sport for old men. Would you agree with that?

5 Tim has shares in three different companies, as shown in the table below.

Share	Number of shares	Total value of shares
FirstBank	400	£3300
Pinwheel	50	£853
PowerCo	2000	£3720

a Which is the cheapest share?

b Pinwheel shares increased in value by $2\frac{1}{4}$%. What is the new value of Tim's Pinwheel shares?

c PowerCo shares decreased in value by 8.2%. What is the value of a single PowerCo share now?

6 This table shows some holiday prices (£) to Tenerife and Gran Canaria.

	Tenerife			Gran Canaria		
Date	S/C	4*HB	A/I	S/C	4*HB	A/I
28 Feb	159	269	335	177	309	319
7 Mar	159	275	339	189	309	319
14 Mar	170	285	349	190	315	335
21 Mar	175	289	349	190	319	335
28 Mar	149	279	339	195	330	339
4 Apr	169	269	345	199	389	379

S/C = self-catering; 4*HB = 4-star hotel, half board; A/I = all-inclusive

a What is the most expensive holiday in March?

b What is the range of prices of holidays?

c From which date is it most expensive to travel to Tenerife?

d Describe any patterns you notice.

e Copy and complete the two-way table below showing the difference in price between holidays to Tenerife and Gran Canaria.

	Difference in price between Tenerife and Gran Canaria		
Date	S/C	4*HB	A/I
28 Feb			
7 Mar			
14 Mar			
21 Mar			
28 Mar			
4 Apr			

f Which type of holiday to Tenerife (S/C, 4*HB or A/I) varies in price the least? Explain your answer.

4.4 Comparing two or more sets of data

1 The chart shows above average rainfall in certain years, which has resulted in increased flooding.

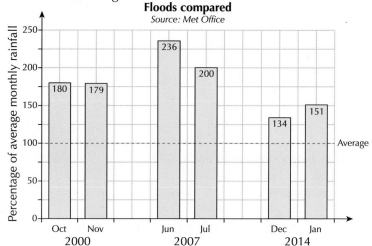

a Which month of which year had the biggest increase from average?
b Explain in words what the 200% increase in July 2007 means.
c There is a slight error in the chart – the final bar is January 2014, so which year does the December bar refer to?
d Find the average increase for each of the three years.
e Round each average to the nearest 10, then express them as a simplified ratio to each other.

2 Answer the questions from the dual bar chart below.

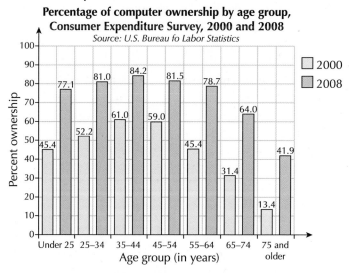

a What was the percentage increase in computer ownership for people aged between 35 and 44 from 2000 to 2008?
b Which age groups went from less than half to over half?
c Which age group more than tripled its ownership?
d Under 25 ownership went up by almost exactly 4% a year on average. Which other age category did this?
e What can you say about the overall trend of computer ownership in these years?

3 The weekly wages for workers in two offices are given below.

Weekly wage (£)	Office A	Office B
Up to 100	2	3
101–200	4	5
201–300	6	4
301–400	5	8
401–500	2	3
501+	1	2

a What fraction of workers in each office receive less than £200 a week?
b What percentage of workers in each office receive over £400 a week?
c What is the most common wage in each office? What is the mathematical term for this?
d Use your previous answers and knowledge of averages to decide which office you would prefer to work in based on the money you could earn. Is this the only thing you should consider when applying for a job at these offices?

 4 People were asked to estimate the number of text messages they sent using their mobile phones in a day. This bar chart shows the results.

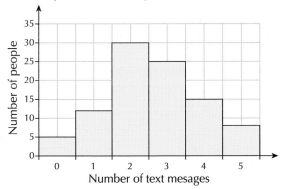

a Five people say they send more than five texts a day so their answers are not included. What category is missing from the chart to allow for this?
b Including these five people, how many people were interviewed in the survey?
c What proportion of these people never text?
d Mervyn says 'Most people send between one and three text messages per day.' Is this true?
e *Mobile Weekly* magazine claim that at least 50% of people send more than three text messages per day. Is this true?
f The same people were asked how many phone calls they make on average each day. The results were as follows:
0 phone calls – 15 people, 1 phone call – 30 people, 2 phone calls – 24 people, 3 phone calls – 18 people, 4 phone calls – 10 people, >4 phone calls – 3 people. Draw a bar chart of these results.
g Daphne claims that more people use the phone at least once than text at least once. Is she correct?

5 The table shows information about the animal populations of four small farms arou
a particular village in the years 2009 and 2014.

Farm	Animals in 2010	Animals in 2015
A	312	220
B	124	165
C	60	72
D	226	144
Total	722	601

a Which farm increased their stock by exactly one-fifth?
b Which farm showed the biggest percentage decrease?
c By rounding the figures to sensible numbers, draw a pie chart for each year. Wh
is the biggest difference visually?
d Draw a dual bar chart for these figures.
e The local newspaper suggests that the size of farms in this area is on the decline
Comment on this suggestion.
f Which of the two charts is better at illustrating your comments about the
newspaper article?

Brainteaser

Although the Football League was formed in 1888, the Premier League began in 1992. These
are the first 20 champions of the Football League along with the Premier League Cup final
results for the same 20-year period.

Season	Champions
1992–93	Manchester United
1993–94	Manchester United
1994–95	Blackburn Rovers
1995–96	Manchester United
1996–97	Manchester United
1997–98	Arsenal
1998–99	Manchester United
1999–2000	Manchester United
2000–01	Manchester United
2001–02	Arsenal
2002–03	Manchester United
2003–04	Arsenal
2004–05	Chelsea
2005–06	Chelsea
2006–07	Manchester United
2007–08	Manchester United
2008–09	Manchester United
2009–10	Chelsea
2010–11	Manchester United
2011–12	Manchester City

Season	Team 1	Score	Team 2
1992–93	Arsenal	1–1	Sheffield Wednesday
1992–93 (Replay)	Arsenal	2–1	Sheffield Wednesday
1993–94	Manchester United	4–0	Chelsea
1994–95	Everton	1–0	Manchester United
1995–96	Manchester United	1–0	Liverpool
1996–97	Chelsea	2–0	Middlesbrough
1997–98	Arsenal	2–0	Newcastle United
1998–99	Manchester United	2–0	Newcastle United
1999–2000	Chelsea	1–0	Aston Villa
2000–01	Liverpool	2–1	Arsenal
2001–02	Arsenal	2–0	Chelsea
2002–03	Arsenal	1–0	Southampton
2003–04	Manchester United	3–0	Millwall
2004–05	Arsenal	0–0	Manchester United
2005–06	Liverpool	3–3	West Ham United
2006–07	Chelsea	1–0	Manchester United
2007–08	Portsmouth	1–0	Cardiff City
2008–09	Chelsea	2–1	Everton
2009–10	Chelsea	1–0	Portsmouth
2010–11	Manchester City	1–0	Stoke City
2011–12	Chelsea	2–1	Liverpool
2012–13	Wigan Athletic	1–0	Manchester City

a   Which teams, if any, won both the League and Cup in the same year?

b   How many of these Cup winners have never won the Premier League?

c   Make a tally of the Cup Final scores. What is the most common result?

d   What percentage of teams failed to score in the Cup Final?

e   What is the mean number of goals scored in these Cup Finals?

f   What fraction of Premier League titles have gone to teams from Manchester?

g   Ethan says Northern teams do better in the League but Southern teams do better in the Cup. Why does he say this?

4.5 Statistical investigations

Investigate one of the following topics by using the information in the suggestions below to complete as many steps as possible.

Step	
1	Decide which general topic to study.
2	Specify in more detail.
3	Consider questions that you could investigate.
4	State your hypotheses (your guesses at what could happen).
5	State your sources of information required.
6	Describe the relevant data.
7	List possible problems.
8	Identify methods of data collection.
9	Decide on the level of accuracy required.
10	Determine sample size.
11	Construct tables for large sets of raw data, in order to make the work manageable.
12	Decide which statistics are most suitable.

1 Tania is investigating our sense of balance. She decides to time how long people can stand on one leg. She thinks that for most people one leg will be better than the other.

2 Darryl is investigating the relationship between diet and exercise. He thinks that vegetarians exercise more than non-vegetarians. He decides to ask people if they are vegetarian and how much they exercise each week. He needs to find enough vegetarians to interview; luckily there is a vegetarian cafe in town.

3 Compare the number of pages in books with the number of chapters.

4 Compare the number of digits that can be accurately remembered by different aged people.

5 Investigate how accurately different people can estimate the weight of household objects. Compare different ages or male and female.

6 Choose an investigation of your own.

5 Applications of graphs

.1 Step graphs

1 An internet provider charges £10 for the first 10 hours spent online. Then for every additional full hour spent online £1.50 is added to the charge.

    a How much does it cost to go online for 7 hours?
    b How much does it cost to go online for $10\frac{1}{2}$ hours?
    c How much does it cost to go online for 12 hours?
    d What is the maximum time that can be spent online for £20?

2 The step graph shows how the cost of a ticket on an underground railway network varies.

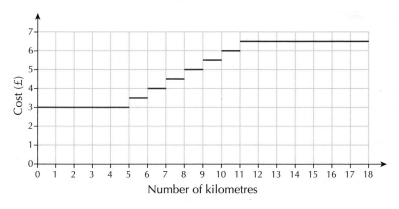

    a What is the cheapest ticket you can buy?
    b What is the fare for a journey of
        i 4.5 km   **ii** 6.5 km   **iii** 8.5 km   **iv** 10.5 km   **v** 12.5 km?

3 The step graph shows how much a library charges in fines for overdue books.

    a What is the longest time you can borrow a book for free?
    b What is the fine for returning a book after
        i four weeks
        ii six weeks and two days
        iii eleven weeks?
    c What is the minimum time a book has been overdue for if the fine is
        i £1.60   **ii** £3.20   **iii** £8?

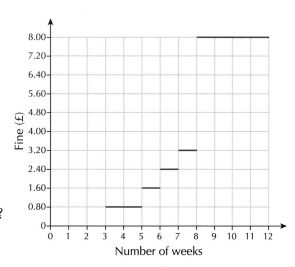

The step graph shows how the taxi fare varies with the distance travelled.

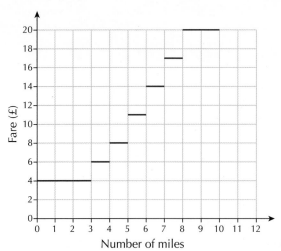

a What is the minimum fare?
b What is the difference in price between
   i a 1 mile journey and a 10 mile journey
   ii a 4.1 mile journey and a 7.9 mile journey?
c What is the furthest you can travel for £11.50?

 A town has two car parks, one for short stays and one for long stays. The step graph shows the different charges at each one.

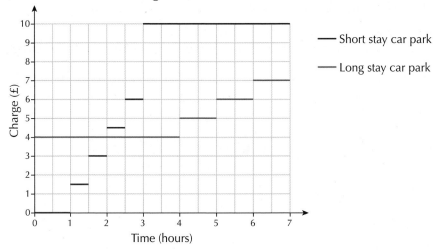

a How much does it cost to park at the short stay car park for
   i 30 minutes   **ii** $2\frac{1}{4}$ hours   **iii** $5\frac{1}{4}$ hours?
b How much does it cost to park at the long stay car park for
   i 30 minutes   **ii** $2\frac{1}{4}$ hours   **iii** $5\frac{1}{4}$ hours?
c Which car park would you recommend to a visitor who wishes to park for
   i 30 minutes   **ii** $2\frac{1}{4}$ hours   **iii** $5\frac{1}{4}$ hours?

6 The step graph shows how much it costs to post a parcel within the same country or abroad.

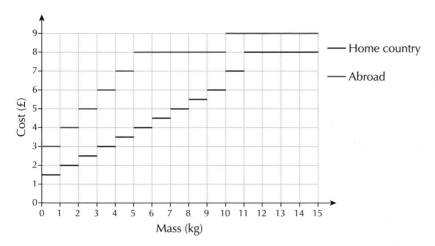

a Anna has a 4.8 kg parcel. What is the difference in price between sending it to her home country and sending it abroad?

b Sarah has £6. How much lighter is the lightest parcel she could send abroad than the heaviest parcel she could send to her home country?

c Grace has two 2.5 kg parcels to send abroad and one 5.7 kg parcel to send to her home country. What is the total cost of postage?

7 A telecommunications company charges 20p for the first minute of a telephone call and then 10p for each subsequent minute or part of a minute.

a How much would a 4 minute 18 second call cost?

b Draw a step graph to show the cost of calls up to 10 minutes in length.

8 Two online retailers charge an extra amount for postage and packaging. The amount charged for postage and packing depends on how much the items in the shopping basket are worth.

Total	Phuntime	Videmus
£0.01–£4.99	£1.50	£0.50
£5.00–£9.99	£2.00	£1.50
£10.00–£16.99	£2.50	£2.50
£17.00–£24.99	£3.00	£3.50
£25.00–£59.99	£3.50	£0.00

a Draw step graphs to show the cost of postage and packaging for Phuntime and Videmus.

b Joseph wants to buy a DVD costing £9.00. Which website would be cheaper?

c Liam wants to buy two DVDs, each costing £9.00. Which website would be cheaper?

d Shayla wants to buy three DVDs, each costing £9.00. Which website would be cheaper?

e Both websites charge £5 for CDs. Which website would you recommend to someone who wants to buy:

   i one CD   ii two CDs   iii three CDs   iv four CDs   v five CDs?

5.2 Time graphs

 1 The graph shows the journeys of two coaches.

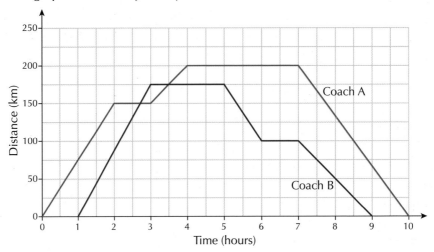

a Which coach travelled furthest?
b Which coach began its journey fastest?
c How long did Coach B stop for in total?

 2 Sketch a graph to illustrate each of the following situations. Estimate any necessary measurements.

a The height of a person from birth to age 30 years.
b The temperature in summer from midnight to midnight the next day.
c The height of water in a WC cistern from before it is flushed to afterwards.

 3 Match each description to its graph.

a The temperature of the desert over a 24-hour period.
b The temperature of a kitchen over a 24-hour period.
c The temperature of a pond during a month of winter.
d The temperature of a cup of tea as it cools down.

i

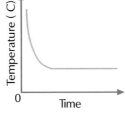

ii

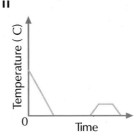

iii

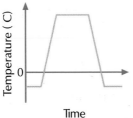

iv

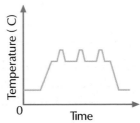

4 George tried several diets to lose weight. The table shows his weight loss over a one-year period. He weighed 105 kg at the beginning of the year.

Draw a graph showing how George's weight changed over time.

Diet	Weight loss/gain
Hay diet	Lost 10 kg in 2 months
Calorie counter	Stayed same weight for 3 months
Vegetarian	Gained 3 kg in 2 months
Usual diet	Gained 12 kg in 1 month
Atkins diet	Lost 28 kg in 4 months

Start the vertical weight axis at 80 kg and use a scale of 1 cm to 2 kg.

5 Ben and Gurjit left their homes at the same time and travelled to each other's home. This graph shows their journeys.

Give reasons for your answers to these questions.

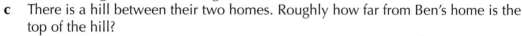

a One person walked and ran, the other cycled. Which person cycled?

b Gurjit and Ben met each other.
  i When was this?
  ii How far were they from Gurjit's home?
  iii How long was their meeting?

c There is a hill between their two homes. Roughly how far from Ben's home is the top of the hill?

d When did Ben travel slowest?

e When did Gurjit travel fastest?

f Ben and Gurjit both started their return journeys at 9.30 am. Ben waited 1 minute at the top of the hill for Gurjit to arrive. They chatted for 2 minutes. Sketch their return journeys.

6 Each of these candles is 25 cm tall and burns to the ground in 20 hours.

For each candle, sketch a graph showing its height over time.

7 Match each graph with the correct description.

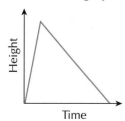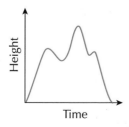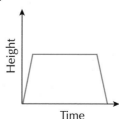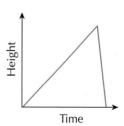

a A hovercraft journey.
b A rocket launch and parachute landing.
c A flight of a radio-controlled model aeroplane.
d A model hot air balloon that takes off and catches fire.

8 Sketch a graph to show these.

    a   The height of a space shuttle that takes off, circles the Earth and lands.

    b   The weight of a woman 3 months before she is pregnant, during her pregnancy and 3 months after she gives birth.

Brainteaser

Emily must catch the 9:00 train from Verlassen to attend a meeting 120 miles away in Ziel. After the meeting she will return home, stopping in Mittagess. The distance–time graph shows her journey.

Ticket prices vary according to distance and are shown in the step graph.

All parts of the journey must be paid for separately.

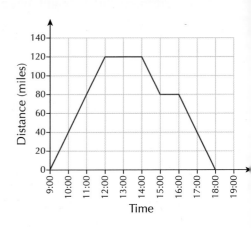

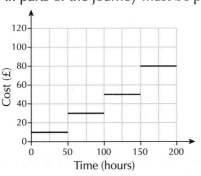

The morning timetable from Verlassen to Ziel is as follows.

Verlassen	8:45	9:00	9:15	9:30	9:45
Mittagess	10:10	10:25	10:40	10:55	11:10
Ziel	11:10	11:25	11:40	11:55	12:10

The afternoon timetable from Ziel to Verlassen is as follows.

Verlassen	14:00	14:15	14:30	14:45	15:00
Mittagess	15:00	15:15	15:30	15:45	16:00
Ziel	16:25	16:40	16:55	17:10	17:25

Any journey that runs more than 30 minutes late qualifies for 5% off the ticket price.

Find the total price Emily must pay for her journey.

5.3 Exponential growth graphs

 1 The table shows the exponential growth of the population of a town.

Year	1990	1995	2000	2005	2010	2015
Population	24 000	36 000	54 000	81 000	121 500	182 250

    a   How many times greater was the population in 1995 than in 1990?

    b   How many times greater was the population in 2000 than in 1995?

    c   Show that the population increased by the same rate every five years.

2 The population growth over time (measured in years) of a village is shown on the graph.

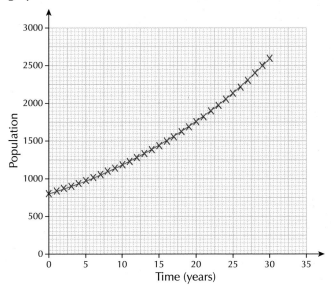

a How many villagers were there initially?
b How many villagers were there after 10 years?
c How many years did it take for the population to reach 2000?
d How much did the population increase by in the first 30 years?

3 The population growth over time (measured in seconds) of a bacteria is shown on the graph.

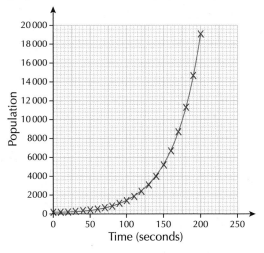

a How many bacteria are alive after 1 minute?
b How many bacteria are alive after $2\frac{1}{2}$ minutes?
c How long did it take for the population to reach 10 000?

4 Bethany invested £3500 in a bank account offering 7% compound interest.

The graph shows how the amount in the account grew exponentially over 50 years.

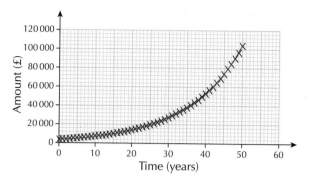

a What was the value of the investment after 20 years?
b What was the value of the investment after 40 years?
c After how many years did Bethany have £80 000 in the account?
d How many years did it take to gain £16 500 interest?

 5 A virus needs to find hosts to infect in order to spread. The graph shows how many hosts have been infected by a particular virus.

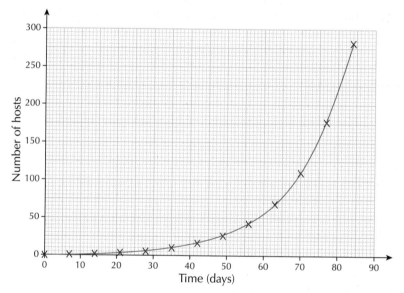

a How many hosts had been infected after three weeks?
b How many hosts had been infected after six weeks?
c How many hosts had been infected after nine weeks?
d How many hosts had been infected after twelve weeks?
e Estimate the number of hosts that will have been infected by the virus after fifteen weeks.
f After how many days had the virus infected at least 100 hosts?

6 The graph shows the temperature of a cup of tea as it cools.

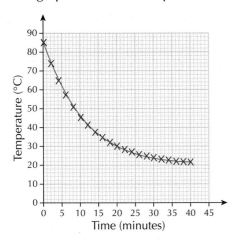

a What was the initial temperature of the cup of tea?
b What was the temperature of the cup of tea after 10 minutes?
c What was the temperature of the cup of tea after half an hour?
d After how many minutes had the tea cooled to 30 °C?
e After how many minutes had the tea cooled by 20 °C?
f The temperature of the tea approaches room temperature as it cools.
Estimate the temperature of the room.

  7 The value of an investment increases exponentially over time.

The table shows the growth of the investment over a three-year period.

Year	2011	2012	2013	2014
Value	£10 400	£10 920	£11 466	£12 039.30

a Draw a graph to show the value of the investment.
b If the investment continues to grow at the same rate, estimate the value of the investment in 2015.

 8 A curve has the exponential equation $y = 3 \times 2^x$.

a Find the value of y when $x = 0$.
b Find the value of y when $x = 3$.
c Find the value of x when $y = 96$.

 9 A curve has the exponential equation $y = ab^x$.

The curve passes through the points with coordinates $(0, 4)$, $(1, 20)$ and $(2, c)$.

a Find the value of a.
b Find the value of b.
c Find the value of c.

6 Pythagoras' theorem

6.1 Introduction

1 a, b and c are integers that represent the three sides of various right-angled triangles. These sets of three integers are known as Pythagorean Triples.

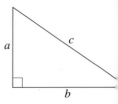

a Copy and complete the table.

a	b	c	a^2	b^2	c^2	$a^2 + b^2$
3	4	5				
6	8				100	
		15	81			225
12				256		

b Which letter represents the hypotenuse?

2 **a** Copy and complete the Pythagorean triples started below.
　i 18, 24, __ 　**ii** 21, __, 35 　**iii** 30, __, __ 　**iv** __, 80, __ 　**v** __, __, 50(
b What ratio do each of these triples simplify to?

(MR) **3** Here are some other Pythagorean triples that follow a different pattern.

5, 12, 13 　　　 7, 24, __ 　　　 9, __, 41 　　　 __, 60, 61

a Find the missing numbers.
b Explain the pattern.

4 Which of these sets of numbers are Pythagorean triples?

　a 1, 2, 3 　　**b** 15, 20, 25 　　**c** 6, 9, 12
　d 20, 32, 45 　**e** 24, 32, 40 　**f** 90, 120, 150

5 Copy and complete the table for $a^2 + b^2 = c^2$.

a	b	c	a^2	b^2	$a^2 + b^2$	c^2
2	3					
10	9					
8		17				
			144	324		

6 Decide whether or not these triangles could exist with the given dimensions.

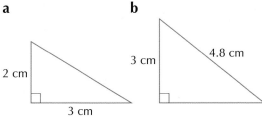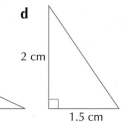

a 2 cm, 3 cm   **b** 3 cm, 4.8 cm   **c** 130 cm, 120 cm   **d** 2 cm, 1.5 cm

 7 Copy this table. Use Pythagoras' Theorem to complete it. Give your answers to one decimal place.

a	b	c	a^2	b^2	a^2+b^2	c^2
1.2	2.3					
5.7						60.8
	12.5				200	
1.5				4		

6.2 Calculating the hypotenuse

 1 Calculate the length of the hypotenuse correct to one decimal place.

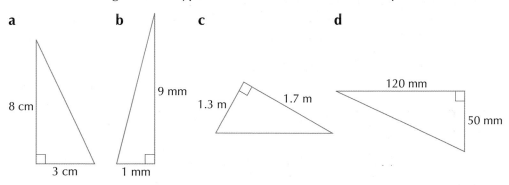

a 8 cm, 3 cm   **b** 9 mm, 1 mm   **c** 1.3 m, 1.7 m   **d** 120 mm, 50 mm

 2 **a** Draw three different right-angled triangles using squared paper.
b Calculate the length of the hypotenuse in each triangle.

3 A helicopter flies 45 km due East, followed by 60 km due North.

a How far is it back to its base in a straight line?
b The aircraft only has enough fuel to cover 200 km.
   What fraction of fuel is left after this journey?

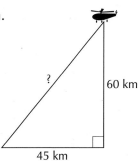

4 Television screens are sized according to the length of their diagonals to the nearest whole number.

  a A TV measures 32 inches across and is 18 inches high. What size would it be classed as?

  Give your answer to the nearest whole number.

  b A TV is advertised as 42 inches, and measures 37 inches across. How high is the screen?

  c The smaller of these two costs £555. How much should the larger one cost to be the same value for money?

5 Tom and James are on a hike. The footpath goes round the edge of a farmer's field, but they decide to cut across the diagonal to save time.

  a How much shorter is it to cut directly across the diagonal?

  b How much time is saved by if they are walking at 5 km/h?

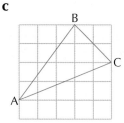

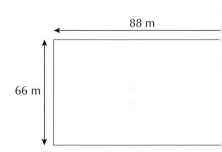

6 Calculate the sides of each of the triangles to the nearest 10 m. Each square represents 100 m.

a

b

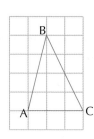

c

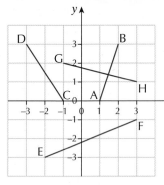

Brainteaser

Use triangles and Pythagoras' theorem to answer the questions about the diagram:

a Find the length of AB to one decimal place.
b Which line measures 5.4 cm to one decimal place?
c Which line is made up of a Pythagorean triple?
d The line GH was intended to be 4.5 cm long to one decimal place. Where should point H have been placed?
e The gradient of a line is the vertical distance divided by horizontal distance.
  i Which line has a gradient of 3?
  ii Which lines have a negative gradient?

1 square = 1 cm

6.3 Calculate the length of the shorter sides

1 Find the length of the missing sides.

a

17 cm

10 cm

b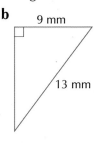

9 mm

13 mm

c

300 mm

200 mm

d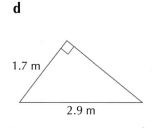

1.7 m

2.9 m

2 A picture in a frame measuring 90 mm by 305 mm is placed diagonally in a box.

The box measures 90 mm by 300 mm. How deep does it need to be for the picture to fit?

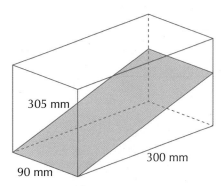

305 mm

90 mm

300 mm

3 This shed is symmetrical and 4 m wide, with a 2.8 m sloping roof. How high is it in total?

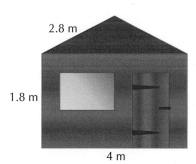

2.8 m

1.8 m

4 m

4 A ladder 4 m long is put against a wall with its foot 2 m away from the wall.

a How far up the wall does it reach?
b The top of the ladder needs to be 20 cm higher. How much closer to the wall does its foot need to be?

5 The maximum allowable gradient for wheelchair ramps is 1:12 as shown.

1 m

12 m

a A community centre's main entrance is 4.5 m above the road level. How long would the ramp have to be to conform to the legislation? Give your answer to one decimal place.
b The centre is only 15 m wide and the ramp cannot be wider than this, so the ramp will have to be in sections. How many sections will the ramp need to be split into?

6 What is the perimeter of this isosceles triangle?

12 cm

x

x

  7 The rectangle is twice as long as it is wide and has the same perimeter as the triangle. What is the area of the rectangle?

a

$2a$

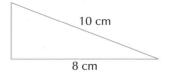

10 cm

8 cm

6.4 Solving problems using Pythagoras' theorem

Give all your answers correct to a suitable degree of accuracy.

1 Calculate the marked lengths.

a

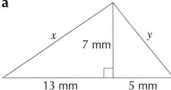

x

7 mm

y

13 mm

5 mm

b

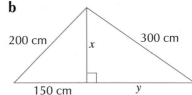

200 cm

x

300 cm

150 cm

y

2 **a** Calculate the length, d, of rope connecting the sailboarder to the kite.

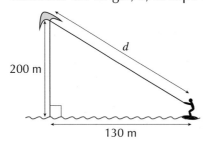

200 m

d

130 m

b Calculate the width, x, of the cellar opening.

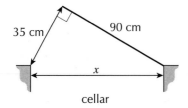

35 cm

90 cm

x

cellar

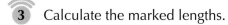

3 Calculate the marked lengths.

a

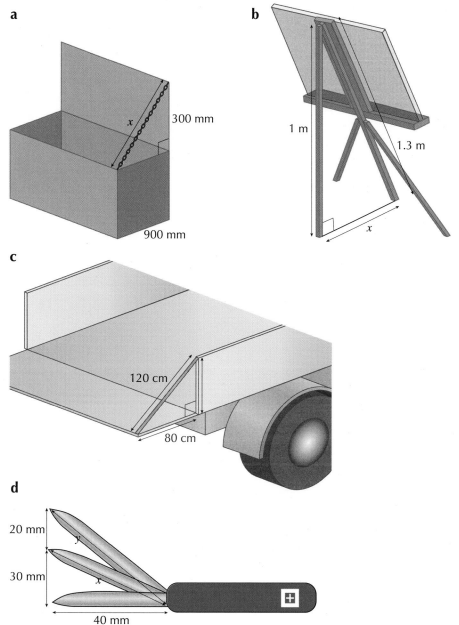

300 mm

x

900 mm

b

1 m

1.3 m

x

c

120 cm

80 cm

d

20 mm

30 mm

y

x

40 mm

4 **a** Calculate the unknown lengths to one decimal place.
   b Explain why triangle ABC is similar to triangle PQR.

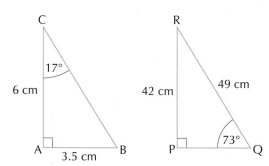

C

17°

6 cm

A

3.5 cm

B

R

49 cm

42 cm

P

73°

Q

5 The diagram shows the **same** folding ladder in different positions. Calculate the marked lengths.

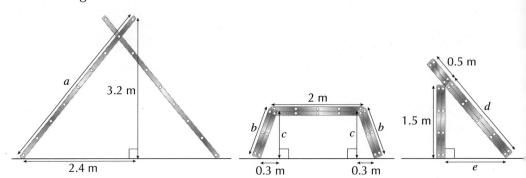

6 The diagram shows the sail of a yacht.

a Explain why triangles PST and PQR are similar.
b Calculate the length of the spar ST.
c How high up the sail is the spar positioned?

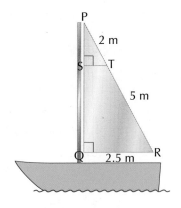

7 Fractions

7.1 Adding and subtracting

1 Find the LCM of each pair of numbers.

    a 3 and 6   **b** 2 and 5   **c** 4 and 12   **d** 12 and 9

2 Use your answers to question **1** to help you to calculate these.

    a $\frac{2}{3} + \frac{1}{6}$ **b** $\frac{1}{5} + \frac{1}{2}$ **c** $\frac{3}{4} - \frac{7}{12}$ **d** $1\frac{1}{12} - \frac{4}{9}$

3 Copy and complete each of the following.

    a $1\frac{2}{3} + 2\frac{1}{3}$ **b** $3\frac{1}{7} + 2\frac{3}{7}$ **c** $\frac{3}{5} + 4\frac{3}{5}$ **d** $1\frac{5}{12} + 6\frac{11}{12}$

    e $3\frac{5}{7} - 1\frac{2}{7}$ **f** $7\frac{4}{11} - 5\frac{3}{11}$ **g** $5\frac{4}{9} - 2\frac{7}{9}$ **h** $8\frac{11}{15} - 3\frac{13}{15}$

4 Convert the fractions to equivalent fractions with a common denominator. Then calculate the answer.

    a $5\frac{1}{8} + 3\frac{1}{4}$ **b** $3\frac{7}{10} + 5\frac{2}{5}$ **c** $1\frac{5}{6} + 2\frac{2}{3}$ **d** $3\frac{9}{10} + \frac{7}{15}$

    e $4\frac{5}{6} - 2\frac{5}{12}$ **f** $6\frac{3}{5} - 4\frac{11}{20}$ **g** $2\frac{11}{12} - \frac{3}{4}$ **h** $9\frac{1}{6} - 4\frac{7}{9}$

5 Marion buys a tub containing $7\frac{7}{8}$ oz of peanuts and a packet containing $3\frac{5}{6}$ oz of peanuts.

    a What is the total weight of the peanuts she bought?
    b How much more does the tub contain than the packet?

6 Calculate these.

    a $2\frac{1}{3} + 1\frac{2}{5}$ **b** $3\frac{1}{4} + 4\frac{2}{3}$ **c** $2\frac{1}{9} + \frac{2}{5}$ **d** $\frac{5}{6} + \frac{17}{20}$

7 Calculate these.

    a $3\frac{3}{4} - 1\frac{2}{3}$ **b** $5\frac{5}{6} - 2\frac{4}{5}$ **c** $4\frac{1}{8} - 1\frac{1}{3}$ **d** $\frac{7}{10} + \frac{9}{20} - \frac{3}{5}$

8 Find the missing fractions.

    a $h + \frac{3}{4} = \frac{5}{6}$ **b** $\frac{7}{8} - m = \frac{3}{10}$ **c** $t - \frac{2}{5} = \frac{2}{3}$

These three paint tins are identical. The fractions show how much paint each tin contains.

$\frac{1}{4}$ full of paint $\frac{2}{5}$ full of paint   empty mixing tin

a What fraction of the tin containing blue paint is empty?
b How much more blue paint than white paint is there?
c A full tin has a capacity of 600 ml.
How much paint is in the blue tin?
d Moira pours the paint from the first two tins into the mixing tin.
What fraction of the mixing tin is filled with paint?
e Altogether, Moira has $2\frac{3}{4}$ tins of white paint and $1\frac{5}{6}$ tins of blue paint.
How many tins of paint does she have altogether?

7.2 Multiplying

1 Work out the following.

  a $\frac{1}{2}$ of 30   **b** $\frac{1}{2} \times 48$ **c** $\frac{1}{3}$ of 27   **d** $\frac{1}{4} \times 24$ **e** $\frac{1}{5}$ of 45

2 **a** Belinda uses $\frac{1}{3}$ of a tin of polish every time she cleans her car.

How many tins has she used after cleaning the car 12 times?

  b Trevor spends one quarter of an hour every day tidying his room.

How many hours does he spend tidying his room in four weeks?

3 Work out the following.

  a $\frac{2}{3} \times 33$ **b** $\frac{2}{3}$ of 27   **c** $\frac{3}{4} \times 24$ **d** $\frac{3}{4}$ of 48   **e** $\frac{2}{5} \times 45$

4 A poster is printed using red, blue and yellow inks. Of the ink used, $\frac{2}{9}$ is red and $\frac{1}{6}$ is blue.

  a To print the poster, 36 ml of ink is used. Calculate the amount of each ink used.
  b Another poster uses 6 ml of red ink. How much ink is used altogether?

5 Match the question cards to the answer cards.

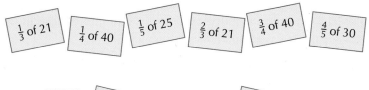

$\frac{1}{3}$ of 21 $\quad$ $\frac{1}{4}$ of 40 $\quad$ $\frac{1}{5}$ of 25 $\quad$ $\frac{2}{3}$ of 21 $\quad$ $\frac{3}{4}$ of 40 $\quad$ $\frac{4}{5}$ of 30

30 $\quad$ 5 $\quad$ 7 $\quad$ 14 $\quad$ 10 $\quad$ 24

  6 Jamie and Ollie are plastering three walls of a rectangular living room. The area is 30 m² in total.

    a If Jamie does $\frac{3}{5}$ of the room while Ollie does the rest, how many square metres does each person do?

    b The boss pays them a total of £90 divided in the ratio of the amount they complete. How much does each person get?

    c The fourth wall in the room is 5 m². What fraction of the room did they plaster?

    d What fraction of the whole room did Jamie plaster?

 7 Work these out.

    a $\frac{4}{9} \times \frac{2}{5}$ **b** $\frac{5}{6} \times \frac{7}{12}$ **c** $\frac{9}{10} \times \frac{7}{11}$ **d** $\left(\frac{2}{5}\right)^2$

 8 Work these out. If possible, cancel before multiplying. Otherwise simplify your answer if necessary.

    a $\frac{5}{8} \times \frac{2}{3}$ **b** $\frac{9}{16} \times \frac{12}{27}$ **c** $\frac{10}{21} \times \frac{7}{15}$ **d** $\frac{8}{9} \times \frac{6}{15} \times \frac{5}{12}$

Brainteaser

In a sports tournament, one player is used as a substitute for the first three matches. In the first game he plays for the final third, then for the last quarter of the second match and the last $\frac{2}{5}$ of the third match. Each match lasts 90 minutes.

a How many minutes did he wait for his first chance to play?
b How many minutes did he play in total? Was it enough to make one complete game?
c What overall fraction of all three games did he play?

7.3 Multiplying with mixed numbers

1 Work these out. Change the mixed numbers to improper (top-heavy) fractions first and simplify your answers, using mixed numbers where necessary.

    a $2\frac{1}{2} \times \frac{1}{8}$ **b** $\frac{1}{5} \times 3\frac{1}{8}$ **c** $4\frac{1}{4} \times \frac{1}{3}$ **d** $\frac{1}{6} \times 3\frac{1}{5}$

    e $\frac{3}{4} \times 6\frac{2}{3}$ **f** $7\frac{3}{5} \times \frac{3}{8}$ **g** $5\frac{3}{11} \times \frac{5}{6}$ **h** $\frac{4}{7} \times 2\frac{5}{8}$

2 In cricket each session usually lasts 2 hours.

  a A player who bats for $2\frac{3}{4}$ sessions has batted for how many minutes?
  b A bowler is used for a quarter of one session, $\frac{2}{3}$ of the next and half of the final one. How many minutes was he bowling for?
  c Due to bad weather one session is extended to last $1\frac{2}{5}$ of its normal time. How long is that?

 3 Konika is doing some design work. She shades all the squares in her first rectangle purple and then part of a second identical rectangle purple.

  a Counting one rectangle as a whole one, what fraction of the two rectangles has been shaded?
  b Konika decides to change a quarter of the shaded squares to yellow. What fraction of a rectangle is yellow?
  c What fraction is still purple?

 4 Write as improper fractions and cancel before multiplying if you can. Otherwise simplify your answer as necessary.

  a $2\frac{1}{4} \times 1\frac{2}{7}$ **b** $2\frac{5}{8} \times 3\frac{5}{9}$ **c** $4\frac{4}{5} \times 3\frac{1}{3}$ **d** $2\frac{7}{10} \times 2\frac{7}{9}$

 5 Work out:

  a $\left(1\frac{3}{4}\right)^2$ **b** $\left(3\frac{1}{2}\right)^2$ **c** $\left(5\frac{3}{8}\right)^2$ **d** $\left(2\frac{1}{3}\right)^2$

 6 Pythagoras' theorem states that $a^2 + b^2 = c^2$. Find c to one decimal place if $a = 2\frac{1}{2}$ an $b = 3\frac{1}{4}$.

7.4 Dividing fractions and mixed numbers

 1 Work out:

  a $5 \div \frac{1}{2}$ $\frac{1}{2} \div 5$

  b $4 \div \frac{1}{3}$ $\frac{1}{3} \div 4$

  c $8 \div \frac{1}{10}$ $\frac{1}{10} \div 8$

2 **a** In question **1**, what did you notice each time?
  b When you divide an integer by a fraction is the answer bigger or smaller?
  c When you divide a fraction by an integer is the answer bigger or smaller?

3 A boxer weighs in at 16 stone which is $\frac{11}{12}$ of the weight of his opponent. How much does his opponent weigh?

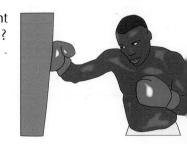

4 Work out:

   a $\dfrac{1}{4} \div \dfrac{1}{3}$ **b** $\dfrac{1}{5} \div \dfrac{1}{8}$ **c** $\dfrac{1}{10} \div \dfrac{1}{6}$

5 Work out:

   a $\dfrac{1}{2} \div \dfrac{3}{5}$ **b** $\dfrac{5}{8} \div \dfrac{2}{3}$ **c** $\dfrac{3}{5} \div \dfrac{7}{10}$

   d $\dfrac{15}{16} \div \dfrac{9}{10}$ **e** $\dfrac{8}{9} \div \dfrac{10}{21}$ **f** $\dfrac{16}{25} \div \dfrac{14}{15}$

6 Ali has just discovered that, at the equator, the Earth is spinning at 1060 mph on average, which is $1\frac{2}{3}$ times faster than London. This is because London is further north and has a smaller circle to travel in the same time.

   a How fast are you spinning if you live in London?
   b Kieran says this should be in km/h, so it needs to be divided by $\frac{5}{8}$. However, Hussain says that's wrong and you should multiply by $1\frac{3}{5}$. Who is right and how fast is London spinning in km/h?

7 Write the numbers as improper (top-heavy) fractions first. Then work out:

   a $2\frac{2}{3} \div 1\frac{2}{5}$ **b** $1\frac{1}{6} \div 2\frac{3}{4}$ **c** $3\frac{5}{9} \div 3\frac{2}{3}$

   d $4\frac{1}{8} \div \frac{33}{40}$ **e** $\frac{8}{15} \div 5\frac{3}{5}$ **f** $2\frac{1}{3} \div 1\frac{8}{9}$

8 A bottle contains $7\frac{7}{8}$ oz of perfume. A perfume spray contains $\frac{7}{10}$ oz of perfume.

   a How many sprays can be filled from the bottle?
   b How much perfume is left over?

Brainteaser

In 2014, the Winter Olympics were held in Sochi, Russia. 2877 athletes took part which was $5\frac{1}{4}$ times the number that entered the Paralympics two weeks later.

a How many Paralympians took part in the games? (Hint: Divide by $5\frac{1}{4}$)

b 85 countries were represented at the 'main' games, which was $1\frac{8}{9}$ times as many as at the Paralympics. How many countries sent athletes to the Paralympic games?

c Why do you think that at the Paralympics more than half the countries were represented, but there were nowhere near half as many athletes?

8 Algebra

8.1 More about brackets

1 Simplify each of these expressions.

   a $9s + 3s$ **b** $6t - t$ **c** $11u + 5u$ **d** $10v - 9v$

   e $8w + w + 7w$ **f** $4x + 5x + 6x$ **g** $9y - 5y + 2y$ **h** $8z - 3z - 2z$

2 Write each expression as simply as possible.

   a $4 \times 2p$ **b** $3p \times 8$ **c** $6p \times 5$ **d** $11 \times 7p$

3 Expand the brackets.

   a $5(d + 2)$ **b** $7(2p - 1)$ **c** $4(3 + m)$ **d** $10(5 - 3i)$

   e $-6(3a + 2)$ **f** $-4(2H - 6)$ **g** $-2(2w + 3)$ **h** $-6(5 - 6x)$

4 Expand the brackets.

   a $a(a + 5)$ **b** $b(4 + b)$ **c** $c(9 - c)$ **d** $d(d + 4)$

   e $3(e + f)$ **f** $g(g - 8)$ **g** $h(h + i)$ **h** $j(4 - j)$

   i $8m(m + 1)$ **j** $3u(4 - u)$ **k** $p(2p + 3)$ **l** $x(5 - 2x)$

   m $i(3i + 2n)$ **n** $-d(2d - 3)$ **o** $-y(5x - 4y)$ **p** $-h(5h + 1)$

5 Check these answers and correct any mistakes.

   a $7x(x + 3) = 7x^2 + 10x$ **b** $x(3x + 5y) = 3x^2 + 5y$ **c** $2t(t - 9) = 2t - 18t$

   d $11m(m - w) = 11m - 11w$

(FS) **6** An envelope costs 5 pence and a stamp costs s pence.

   a Write down an expression for the cost of:

     i a stamped envelope

     ii 6 stamped envelopes

     iii n stamped envelopes.

   b Expand the brackets for your answers to parts **ii** and **iii**.

7 A box contains b grams of salt. The box weighs 10 g.

   a Write down an expression for the total weight of these:

     i a box of salt

     ii 3 boxes of salt

     iii b boxes of salt.

   b Expand the brackets for your answers to parts **ii** and **iii**.

(MR) **8** **a** Find an expression for the missing length.

   b Find two different expressions for the shaded area.

   c Show that your expressions in part **b** are equivalent.

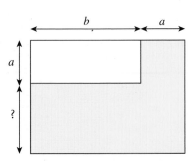

9 Expand the brackets and simplify as much as possible.

    a $g(g-3)+7g$ **b** $g(4+g)-5g$ **c** $g(2g-9)+g^2$ **d** $g(12-g)-8g$

10 Simplify each expression.

    a $(5q)^2$ **b** $(7q)^2$ **c** $(11q)^2$

11 Expand the brackets.

    a $3x(3x+5)$ **b** $4x(2x^2+9)$ **c** $5x^2(11-x)$ **d** $6x^2(10x-7)$

12 Simplify each expression.

    a $(3t)^3$ **b** $(4u^2)^3$ **c** $(2v^6)^5$

13 Expand the brackets.

    a $2y^3(7y^2-4)$ **b** $8y^7(6+9y^5)$

8.2 Factorising expressions containing powers

1 Factorise each expression as much as possible.

    a $5m+20$ **b** $12x-18$ **c** $30+8k$ **d** $14-21h$

2 Factorise each expression as much as possible.

    a $3m-3n$ **b** $16p+8q$ **c** $27a-18b$ **d** $-20f-10g$

3 Factorise each expression as much as possible.

    a $ab+2a$ **b** $mn-m$ **c** $2st+5s$ **d** $de+df$

    e $4ij+4i$ **f** $9r-6rw$ **g** $15pq+20pr$ **h** $6hu+12iu$

4 Complete each factorisation.

    a $w^2-8w=w(...)$ **b** $x^2+5x=x(...)$ **c** $y^2-13y=y(...)$ **d** $2z-z^2=z(...)$

5 Factorise each expression as much as possible.

    a d^2+3d **b** $4t-t^2$ **c** y^2-9y **d** $i+6i^2$

    e p^2-p **f** $2d+d^2$

6 Some of these expressions can be factorised and some cannot.

    Factorise them if you can. Write 'Not possible' if they can't be factorised.

    a $5a+a^2$ **b** $3b-10$ **c** $4c^2-5c$

    d $4d+12$ **e** $6e^2+1$ **f** $8f^2-24$

    g $8g-g^2$ **h** $7h-14$ **i** i^2+10j

7 **a** Find the value of $4u^2+u$ when $u=5$.

    b Factorise $4u^2+u$.

    c Show that you get the same value as part **a** when you substitute $u=5$ into the factorisation from part **b**.

8 Factorise each expression as much as possible.

   a $5u^2 + 5u$ **b** $3e^2 - 9e$ **c** $8p^2 + 12p$ **d** $12gh + 6h$

   e $10w^2 - 2w$ **f** $7cd + 4c$ **g** $10n - 25n^2$ **h** $-7m^2 - 14m$

 9 Factorise each expression as much as possible.

   a $k^3 + 6k$ **b** $4k^3 + 8k$ **c** $10k + 15k^3$

   d $k^3 + 2k^2$ **e** $8k^2 - 6k^3$

10 Factorise each expression as much as possible.

   a $8b^2 + 12b^5$ **b** $14b^6 - 7b$ **c** $35b^{11} + 25b^7$ **d** $42b^{42} + 49b^{49}$

Brainteaser

x, $(x + 1)$ and $(x + 2)$ are three consecutive integers.

a What is the sum of the three consecutive integers?

b Factorise your answer to part **a**.

c Use your factorisation to prove that the sum of three consecutive integers is always a multiple of 3.

d Is the sum of four consecutive integers always a multiple of 4?

e Is the sum of five consecutive integers always a multiple of 5?

f What can you say about the sum of n consecutive integers?

8.3 Expanding the product of two brackets

1 Expand the brackets.

   a $3(x + 5)$ **b** $2(x - 9)$ **c** $4(x + 8)$

   d $6(3 - x)$ **e** $10(2x + 7)$ **f** $7(5 - 3x)$

 2 Expand the brackets.

   a $(d + 5)(e + 6)$ **b** $(u + 3)(v + 10)$ **c** $(y + 2)(z + 8)$

   d $(p + 5)(q - 7)$ **e** $(p - 5)(q + 7)$ **f** $(p - 5)(q - 7)$

 (MR) **3** **a** Explain why the area of the blue rectangle can be written as $(s + t)(u - v)$.

   b Use the diagram to explain why the area of the blue rectangle can also be written as $su + tu - sv - tv$.

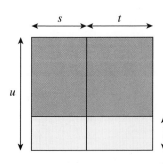

 (MR) **4** **a** Find the value of 13×12.

   b Expand and simplify $(x + 3)(x + 2)$.

   c Substitute $x = 10$ into your answer to part **b**.

   d Your answers to parts **a** and **c** should be the same. Explain why.

5 Expand the brackets. Simplify the result by collecting like terms.

   a $(x+3)(x+1)$ **b** $(x+9)(x+2)$ **c** $(x+5)(x+4)$

   d $(x-3)(x-5)$ **e** $(x-1)(x-10)$ **f** $(x+5)(x-2)$

   g $(x-7)(4+x)$ **h** $(x-6)(x-4)$ **i** $(x-8)(x+9)$

6 **a** Expand the brackets. Simplify the result by collecting like terms.

      i $(c+5)(c-5)$ **ii** $(f+4)(f-4)$ **iii** $(t+9)(t-9)$

   b Explain why each answer has only two terms.

7 Fiona's homework has been smudged. Fill in the gaps.

   a $(a+5)(a-2) = a^2 + \ldots\ldots - 10$

   b $(b+7)(b+8) = b^2 + 15b + \ldots\ldots$

   c $(c-3)(c-7) = c^2 - \ldots\ldots + \ldots\ldots$

   d $(4+d)(9+d) = d^2 + \ldots\ldots + 36$

   e $(e+11)(e+5) = \ldots\ldots + 16e + \ldots\ldots$

   f $(f-9)(f+1) = f^2 - \ldots\ldots - \ldots\ldots$

8 Explain why $(x+3)^2 = x^2 + 6x + 9$.

9 Expand and simplify:

   a $(x+5)^2$ **b** $(x+4)^2$ **c** $(x-7)^2$

   d $(x-8)^2$ **e** $(x+100)^2$ **f** $(x-200)^2$

9 Decimal numbers

9.1 Powers of 10

1 Multiply each of the following numbers by: **i** 10 **ii** 100 **iii** 1000

   a 7   **b** 84   **c** 273   **d** 950   **e** 5800
   f 0.42   **g** 874   **h** 12.6   **i** 0.053   **j** 0.0004

2 Divide each of the following numbers by: **i** 10 **ii** 100 **iii** 1000

   a 8000   **b** 390 000   **c** 4 639 000   **d** 58 000   **e** 82 000 000
   f 4300   **g** 0.6   **h** 23.7   **i** 0.054   **j** 13 599

3 Multiply these numbers by: **i** 10^3 **ii** 10^7 **iii** 10^{-1} **iv** 10^{-3}

   a 600   **b** 5   **c** 23 000   **d** 0.6   **e** 25.2

(MR)

4 Explain why each of these sums gives the same answer.

   35×0.01, 35×10^{-2}, $35 \div 100$, $35 \div 10^2$

5 Calculate each of these.

   a 5.8×10^4 **b** 0.33×10^3 **c** $51.188 \div 10^3$ **d** 0.095×10^{10}
   e 7.1×10^8 **f** $0.2 \div 10^5$ **g** $23.4 \div 10^1$ **h** 11.9×10^{-4}

6 In the 2011 census, the population of Cornwall was estimated to be 532 300.

   The area of Cornwall is 3563 square kilometres.
   a What is the average land area per person in Cornwall?
   b What is the average number of people per square kilometre in Cornwall?
   Give both answers correct to two significant figures.

7 In 2012, the population of Brazil was approximately 199 million people.

   Brazil covers an area of approximately 8.5 million square kilometres.

   What is the average number of people per square kilometre in Brazil, correct to two significant figures?

8 Put these European countries in order from least densely populated to most densely populated.

Country	Area	Population (2012)
Switzerland	41 285 km²	8.0 million
Netherlands	41 526 km²	16.8 million
Portugal	91 985 km²	10.5 million
France	674 843 km²	65.7 million
Belgium	30 528 km²	11.1 million

9.2 Standard form

1. Which of these numbers are written in standard form?

   a 40×10^3 b 2.3×10^7 c 4×10^{-3}
   d 1×10^{10} e 10^9 f 0.8×10^5

2. Write these as ordinary numbers.

   a 2.1×10^3 b 4×10^8 c 3.9×10^6 d 9.9×10^9
   e 7.9×10^{-2} f 4×10^{-7} g 4.6×10^{-4} h 1.8×10^{-5}

3. Write these numbers using standard form.

   a 3200 b $900\,000$ c 420 d $8\,000\,000\,000$
   e 0.054 f $0.000\,007$ g 0.3 h 0.0044
   i 27×10^2 j 0.07×10^7 k 400×10^{-5} l 0.13×10^{-2}

4. Find the square of each number, giving your answer in standard form.

   a 5 b 600 c 0.003 d 0.00012

5. Write each answer in standard form, correct to two significant figures.

   a $6.5 \times 10^{10} + 3.4 \times 10^{10}$ b $6.4 \times 10^4 + 2.8 \times 10^3$
   c $2.525 \times 10^9 + 3.131 \times 10^8$ d $8.3 \times 10^7 - 4.6 \times 10^6$
   e $4.3 \times 10^{-1} + 2.8 \times 10^{-1}$ f $3.7 \times 10^{-2} - 7.3 \times 10^{-3}$

 (MR) 6. Explain why $3 \times 10^8 + 2 \times 10^7$ does not equal 5×10^{15}.

7. Write each answer in standard form, correct to 2 significant figures.

   a $7.5 \times 10^{12} + 3.2 \times 10^{12}$ b $9.237 \times 10^8 - 8.618 \times 10^8$

8. The mass of an electron is $0.000\,000\,000\,000\,000\,000\,000\,000\,000\,000\,91$ kg.

   a Write the mass of an electron in grams. Give your answer in standard form.
   b Write the mass of two electrons in grams. Give your answer in standard form.
   c How many electrons are there in 1 gram? Give your answer in standard form, correct to two significant figures.

9. Calculate these. Give your answers in standard form. Do not use a calculator. Show your working.

   a $(4 \times 10^3) \times (2 \times 10^2)$ b $(7 \times 10^4) \times (4 \times 10^5)$
   c $(4 \times 10^6) \div (2 \times 10^2)$ d $(7 \times 10^4) \div (4 \times 10^6)$

10 A digital photo has a size of 1.2×10^6 bytes. How many photos can be stored on a computer hard drive with a capacity of 2.8×10^{11} bytes?

 11 Light travels at a speed of 2.99×10^8 metres per second. The distance of the Earth from the Sun is 1.526×10^8 km. How long does it take for light to travel from the Sun to the Earth?

Brainteaser

A googol is 10^{100}.

Write each of these in standard form:

a 52 googols
b Six thousand eight hundred googols
c Seven million googols
d One-quarter of a googol
e 3% of a googol
f 43 millionths of a googol
g The square of a googol
h 80 googols squared
i 3 googols cubed
j Two divided by a googol

9.3 Rounding appropriately

 1 Round each number to
i one decimal place **ii** two decimal places.

a 6.715	**b** 3.953	**c** 4.288	**d** 7.035	**e** 1.899
f 0.05537	**g** 20.957	**h** 0.5059	**i** 334.58943	

 2 Which measurement has been rounded to the most appropriate degree of accuracy? Explain your answer.

a The mass of a bag of potatoes
  i 202.45236 g **ii** 202.5 g **iii** 200 g

b The maximum speed of a car
  i 150 mph **ii** 146 mph **iii** 100 mph

3 Round each number to the nearest whole number and use these answers to estimate the value of each calculation.

    a $6.2 + 8.7$ **b** $9.4 - 2.9$ **c** 4.2×8.9 **d** $9.6 \div 2.1$

4 Round each number to the nearest 10 and use these answers to estimate the value of each calculation.

    a $43 + 91$ **b** $62 - 28$ **c** 51×78 **d** $76 \div 18$

5 Round each number to:

    i   one significant figure   **ii**   two significant figures.

    a   6.715   **b**   3.953   **c**   4.288   **d**   7.035   **e**   1.899

    f   0.05537   **g**   20.957   **h**   0.5059   **i**   334.58943

6 Estimate answers to these by first rounding the numbers to one significant figure.

    a 0.76×47 **b**   12% of \$2761   **c** $38\,297 \div 77$ **d** $5.38(13.85 - 4.199)$

    e $35.44^2 - 22.2^2$ **f** $0.9167 \div 0.0286$ **g** $\dfrac{5.018 \times 3.86}{11.41 - 6.459}$ **h** $\dfrac{37.71 + 18.099}{0.829 - 0.333}$

7 Round these quantities to an appropriate degree of accuracy.

    a   Jeremy is 1.8256 m tall.
    b   A computer disk holds 688 332 800 bytes of information.
    c   A petrol tank has a capacity of 72.892 litres.
    d   An English dictionary contains 59 238 722 words.

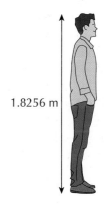

1.8256 m

    e   Kailash held his breath for 82.71 seconds.

8 Use a calculator to work out these.

Round your answers to a suitable degree of accuracy.

    a 7392^2 **b** $\dfrac{5}{7}$ **c** $18.3 + 2.8 \times 9.7 \times 1.4$ **d** $\dfrac{997 - 285}{0.863 \times 5.25}$

9 Grace, Victoria and Zeenat all measured the length of a boat to a different degree of accuracy.

Grace measured the length as 20 m.

Victoria measured the length as 18 m.

Zeenat measured the length as 18.5 m.

    a   If all three were correct, write down a possible length for the boat.
    b   Write down what rounding each person did.

10 This table shows the prices of some metals.

Metal	Price
lead	£0.2412 per kg
silver	£0.085 per g
gold	£2575 per kg
platinum	£11.63 per g

a Estimate the cost of buying these. Show your working.

   i 2474 kg of lead   **ii** 0.2856 g of silver
   iii 0.036 kg of gold   **iv** 19.08 g of platinum

b Estimate answers to these. Work in pounds. Show your working.

   i How much gold could be bought for £64 250?
   ii How much lead could be bought for £124.59?
   iii How much platinum could be bought for £289 320?
   iv How much silver could be bought for £0.94?

9.4 Mental calculations

Do not use a calculator to answer any of these questions.

 1 Work these out mentally.

   a 7 × £2.99   **b** 11 × £3.98   **c** 6 × £4.90

2 Bethany bought seven DVDs. Each DVD cost £6.98.
How much change should Bethany get from a £50 note?

**All DVDs
£6.98 each**

 3 Work out each of these in your head and write down the answer.

a 800 × 25	**b** 164 × 25	**c** 350 ÷ 25	**d** 31 ÷ 25
e 14 × 15	**f** 32 × 15	**g** 1460 ÷ 20	**h** 26 480 ÷ 20
i 4000 ÷ 16	**j** 960 ÷ 16		

 4 Work out the answer to each of these. Use the method you like best.

a 8000 × 25	**b** 16 400 × 25	**c** 35 ÷ 25	**d** 31 000 ÷ 25
e 140 × 15	**f** 3200 × 15	**g** 14.6 ÷ 20	**h** 2 648 000 ÷ 20
i 400 000 ÷ 16	**j** 0.96 ÷ 16		

5 Work out the answer to each of these. Use the method you like best.

a 8000 × 2.5	**b** 16 400 × 0.25	**c** 35 ÷ 250	**d** 31 000 ÷ 2.5
e 140 × 1.5	**f** 3200 × 150	**g** 14.6 ÷ 2000	**h** 2 648 000 ÷ 20
i 400 000 ÷ 160	**j** 0.96 ÷ 0.16		

6 Using suitable rounding, estimate each of these in your head. Write down the estimated answer.

a 1199×24 b $8641 \div 19.6$ c $6012 \div 16.03$

d $20\,202\,020 \times 14.7$ e $111 \div 25.2$

7 Work these answers out in your head. Then write down the answers.

a A block of platinum weighs 8400 kg. What would $\frac{1}{16}$ weigh?

b What is the cost of 17 pads at 99p each?

c A keycutter charges £2.40 to repair a pair of shoes. How much would she charge to repair 15 pairs of shoes?

d Valerie has 860 lottery tickets to sell for a charity event. If she only sells $\frac{1}{20}$ of them, how many is that?

8 A rectangle has an area of 1203 cm².

Its base is 24.7 cm.

Find an estimate for the perimeter of the rectangle.

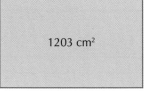

1203 cm²

24.7 cm

9 One euro is approximately 14.48 South African Rands.

Maduraa has €24.96.

She wants to buy three bracelets, each costing 119 South African Rands.

Use estimates to work out if Maduraa has enough money.

10 Find, without using a calculator:

a 888×0.025

b $2000 \div 1.6$

c 42×150

11 Venus' diameter is about 7521 miles.

The diameter of Mercury is about 2.5 times smaller.

Find an estimate for the diameter of Mercury.

12 A theatre makes £28 profit for each ticket sold for a musical. If 391 tickets are sold, find the approximate profit.

Brainteaser

A cuboid block measures 3.1267 m by 3.9238 m by 5.9756 m.

Harriet has to paint 25 identical cuboid blocks.

Each can of paint has enough to coat 19.8 m².

She has 120 cans of paint.

Without using a calculator, estimate whether or not she has enough paint.

9.5 Solving problems

  1 Rita and Norris work in a shop. Norris works twice as many hours as Rita.

Altogether they work 48 hours. How many hours does Rita work?

  2 The total capacity of five cups and three mugs is 230 cl.
The total capacity of five cups and four mugs is 265 cl.

   a Find the capacity of a mug.
   b Find the capacity of a cup.

  3 **a** Tamsin converted £200 to euros and then bought some fruit juice.
How many euros did she receive?

   b She bought as many bottles of fruit juice as possible.
How much fruit juice did she buy?

   c **i** How much money did she have left over?

     ii Convert your answer to British pounds.

Exchange rates
£1 = €1.48

€3.50 per bottle

4 Which biscuit purchase gives the best value? Explain your answer.

  5 The TV programme Wacko Magic lasts 25 minutes and is shown every Tuesday and Thursday. Country Facts lasts 15 minutes and is shown every Wednesday. Karen records these programmes each week. She has recorded 7 hours and 35 minutes on a videotape.

   a How many weeks did she record?

   b How many of each programme did she record?

  6 Mr Marshall has two children, Daniel aged 9 and Kate aged 12. He divided £735 between them in the ratio of their ages. The children bought a computer between them for £420. Kate paid twice as much as Daniel.

What is the ratio of the money the children are left with?

  7 Tea bags are sold in boxes of 200, 300 and 500.

A box of 200 costs £1.72.

A box of 300 costs £2.61.

A box of 500 costs £4.45.

Which box is the best value for money? Explain your answer.

  8 Tribonacci sequences are formed by adding the previous three numbers to get the next number.

   a Write down the next three terms of the tribonacci sequence that starts: 1, 1, 1, …

   b Work out the missing values of this sequence: …, …, …, 4, …, 7, 16, …, …, …, 174

   c Explain why all the terms (apart from the first two) in this tribonacci sequence end in a 1 or a 9: 3, 7, 11, ….

10 Prisms and cylinders

10.1 Metric units for area and volume

1 Convert these measurements to square centimetres.

    a $5\,m^2$ **b** $0.07\,m^2$ **c** $8340\,mm^2$ **d** $16\,mm^2$

2 Convert these measurements to square metres.

    a $3000\,cm^2$ **b** $12\,500\,mm^2$ **c** $13\,499\,cm^2$ **d** $8\,km^2$

3 **a** A hockey pitch is 80 m long and 45 m wide. Work out
its area in:
    i square metres   **ii** hectares.

    b A garden path is made from 350 rectangular paving
stones each 30 cm long and 15 cm wide.
How many square metres does it cover?

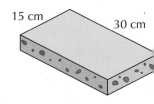

4 Convert these measurements to cubic centimetres.

    a $2\,m^3$ **b** $0.5\,m^3$ **c** $78\,mm^3$ **d** $9300\,mm^3$

5 Convert these measurements to cubic metres.

    a $8\,000\,000\,cm^3$ **b** $675\,cm^3$ **c** $25\,000\,000\,mm^3$

6 Convert these quantities (remember that cubic centimetres are the same as cm^3).

    a $8400\,cm^3$ to litres   **b** 65 cubic centimetres to litres **c** 4.8 ml to cm^3
    d 200 ml to litres   **e** 9 litres to cm^3 **f** 3.75 litres to ml
    g $1.5\,m^3$ to litres   **h** 275 litres to m^3

7 **a** How many square metres is $\frac{1}{4}$ of a hectare?
    b What fraction of a square kilometre is 2500 square metres?
    c Express 35 cubic centimetres as a percentage of a litre.

8 A doctor prescribes some medicine for a patient with instructions to
take two 5 ml spoonfuls three times a day for seven days. This uses up
the whole bottle. How many cubic centimetres are in the bottle?
What percentage of a litre is this?

9 **a** In cubic centimetres (cc), how much capacity does a car with a
2.3 litre engine have?
    b A car engine has a cylinder capacity of 1487 cc but when advertised this is
rounded to the nearest 100 and converted to litres. What size will it be
advertised as?

Brainteaser

a How much more coke do you get from a 2 litre bottle than six cans of 330 ml?

b What is this as a percentage?

c If the cans cost £1.98 for six, how much should the bottle cost to be of equal value?

d One week, two extra free cans are offered for £2.30. How many litres is that altogether?

e Is this better value than a 2.5 litre bottle priced at £2.25?

10.2 Volume of prisms

1 Calculate the volume of each cuboid in the most appropriate units.

a   300 cm   200 cm   400 cm

b   20 cm   50 cm   30 cm

c   6 cm   6 cm   125 cm

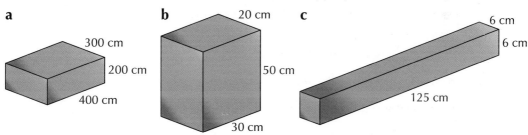

2 Calculate the capacity (in litres or ml as appropriate) of each cuboid in question **1**.

3 **a** Calculate the volume and capacity of a cube of side 15 cm.
　 b A cube has a volume of 8000 cm³. What are the lengths of its sides?

4 The picture shows a 3 litre tin of biscuits. The area of the lid is 375 cm². How deep is the tin?

PS **5** A water tank is in the shape of a trapezium as shown. An overflow pipe is fixed 20 cm below the top of the tank. What is the greatest volume of water the tank can hold?

110 cm

90 cm

120 cm

70 cm

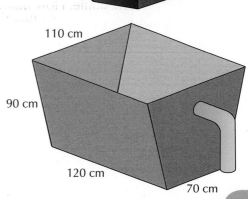

FS
PS

6 The box below is filled with packets of tea. The packets are 5 cm square and 12 cm high. When full, the box contains 8 dozen packets of tea which are stacked upright in the box.

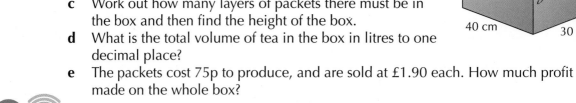

a What volume of tea is in each packet, in millilitres?
b How many packets of tea fit into the base of the box?
c Work out how many layers of packets there must be in the box and then find the height of the box.
d What is the total volume of tea in the box in litres to one decimal place?
e The packets cost 75p to produce, and are sold at £1.90 each. How much profit is made on the whole box?

40 cm   30 c

FS

MR

7 Krispies are sold in three sizes: mini, medium and giant. The boxes are filled to the top.

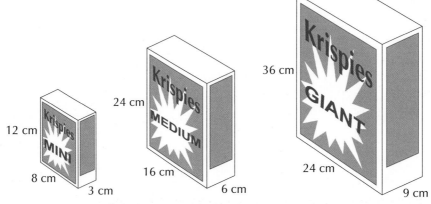

12 cm   24 cm   36 cm
8 cm   16 cm   24 cm
3 cm   6 cm   9 cm

a Write the ratio of the heights of the three boxes in its simplest form.
b Calculate the volume of each box.
c Write the volumes as a ratio in its simplest form.
d Can you see a link between the ratios in parts **a** and **c**?
e How many mini boxes could be fitted into:
  i a medium box   **ii** a giant box?
f The answers in part **e** are the same as which other answers you have found? Why is that?
g The prices of these packets are 45p, £2.70 and £6.75 respectively. Express these as a simple ratio.
h Why do you think the ratio of the prices is different to the ratio of the volumes?

Brainteaser

In 2017 the government intends to introduce a new 12-sided £1 coin as shown below.

a What is the mathematical name of a 12-sided shape?
b The old £1 coin is a circle with diameter 22.5 mm and is 3.15 mm thick. Find its volume to the nearest cubic millimetre.
c The new coin is intended to be as close as possible in volume to the old one. It is made of 12 triangles, each with an area of 30 mm^2. How thick would it need to be to the nearest hundredth of a millimetre?
d Even though they are different sizes, the density (mass ÷ volume) of both a £1 coin and a £2 coin are identical. The £1 coin has a mass of 9.5 g, whereas the £2 coin weighs 12 g and is 2.5 mm thick. What is the diameter of a £2 coin in millimetres, to one decimal place?

10.3 Surface area of prisms

1 Calculate the surface area of each prism.

a

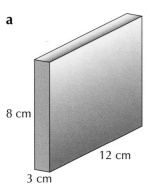

8 cm

12 cm

3 cm

b

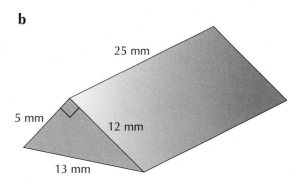

25 mm

5 mm

12 mm

13 mm

2 Convert your answers to question **1** to square metres.

3 A fish tank with glass walls has dimensions 60 cm by 30 cm by 20 cm with a solid base and no top. How much glass is needed?

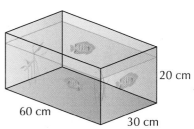

20 cm

60 cm

30 cm

4 The diagram shows a petrol tank in the shape of a prism.

    a Calculate the area of the cross-section.
    b Calculate the volume of the tank.
    c Calculate the capacity of the tank in litres.
    d Calculate the surface area of the tank in square metres.

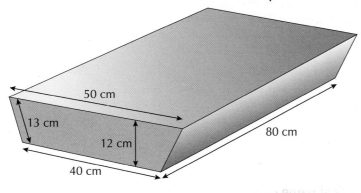

50 cm

13 cm

12 cm

80 cm

40 cm

5 Look at the dimensions of the Krispies packets carefully.

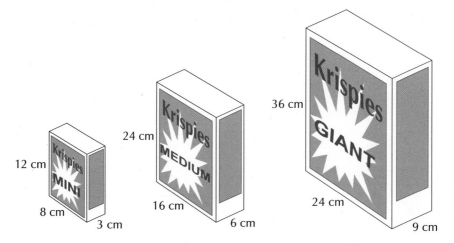

a Work out the surface area of each packet.

b Write these areas as a ratio in its simplest form.

c There is a quick way to simplify their ratios from their dimensions. Explain the method.

d Are these boxes similar shapes or not? Explain your answer.

PS **6** Tom has a small tent which is 1 m wide, 1.2 m high, 2.4 m long and has a sewn-in groundsheet.

a Use Pythagoras' theorem to find the sloping height of the tent.

b Use your previous answer to work out how much material is needed to make this tent.

When Tom camps with Scouts they have bigger tents that allow more height before the sloping roof begins, as shown in the second diagram. The extra vertical bits of canvas are 50 cm high, joining to a sloping section of 1.5 m. They are 1.8 m wide and 2.8 m long, but have no groundsheets.

c Work out the vertical height of these tents.

d How much material is used to make these tents?

e How much extra space do you get inside the scout tent compared to Tom's own tent?

Brainteaser

a Use Pythagoras' Theorem to work out the area of the triangle and hexagon in the shapes below:

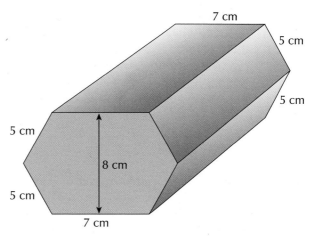

b Both prisms have the same volume. If the triangular prism is 12 cm high, how long must the other prism be?
c If the hexagonal prism is 15 cm long, how high must the triangular prism be?
d Use ratios to find a quick way to find the height or length of one shape if you know the height or length of the other.
e Use algebra to show that the surface areas are the same when the hexagonal prism is $2\frac{6}{7}$ cm long.

10.4 Volume of a cylinder

1 Work out the volume of these cylinders:

    a cross-sectional area 3 cm², height 15 cm
    b cross-sectional area 1.5 m², length 4 m
    c cross-sectional area 5.4 cm², width 8 mm

2 Find the volume of these cylinders. Give your answers to one decimal place.
    a radius 4 cm, length 10 cm
    b radius 2.5 m, height 6 m
    c diameter 12 cm, length 15 cm
    d diameter 8.4 m, height 7.5 m

3 A circular pool has a radius of 1.5 m and a depth of 60 cm.

What is the capacity of the pool to the nearest 100 litres?

How much water does it contain when $\frac{2}{3}$ full?

4
  a If this cylinder has a volume of 340 cm^3 and a radius of 3 cm, what is its height to the nearest centimetre?
  b If the volume is 37 m^3 and it is 6 m high, what is its radius to one decimal place?
  c If it has a capacity of 2 litres, with a diameter of 8 cm, how high will it be to the nearest centimetre?

(PS) **5** A water pipe has an internal diameter of 40 cm.
  a Calculate:
   i its radius
   ii the cross-sectional area
   iii the volume per metre of the pipe
   iv the quantity of water, to the nearest litre, in every metre of the pipe.
  b If the water in the pipe moves at a speed of 0.8 metres per second, how many litres, to the nearest 1000, will pass through it in one hour?

(PS) **6** A component is manufactured by cutting a circular hole of radius 6 mm through a metal block.

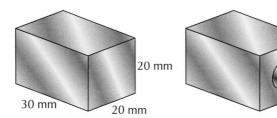

  a Find the volume of the block after the hole has been drilled.
  b What percentage of the block has been removed? What is this to the nearest unit fraction?
  c The original block weighs 252 g. What does it weigh after the hole is removed?

7 The cross-section of this plastic bench has an area of 620 cm^2. Calculate its volume in
  a cubic centimetres   **b** cubic metres.

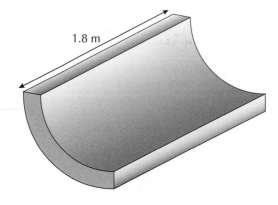

1.8 m

8 The picture shows an urn containing hot water. It has an internal diameter of 24 cm and is 40 cm high. The serving tap is located 5 cm above the base, and once the water reaches this level, no more will come out.

a Calculate the total capacity of the urn to the nearest millilitre.

b Calculate the usable amount of water in the urn to the nearest millilitre.

c What percentage of the total capacity is usable?

d In the canteen where it is used, tea mugs hold about 25 cl, but coffee is served in special 20 cl cups.

If an equal number of both drinks are served, what is the maximum number of cups that can be filled?

(Hint: Use algebra if you can.)

10.5 Surface area of cylinders

1 Find the surface areas of these cylinders.

a radius 4 cm, length 10 cm
b radius 2.5 m, height 6 m
c diameter 12 cm, length 15 cm
d diameter 8.4 m, height 7.5 m

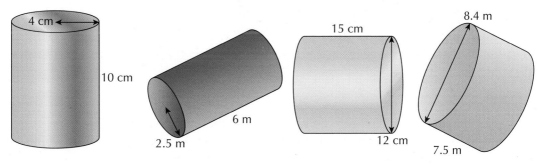

2 A pond has a surface area of 20 m^2. What is its radius to the nearest centimetre?

3 A trampoline with a diameter of 12 m has protective netting all the way round to a height of 2 m. What area of netting is required?

4 A circular sand pit is to be built in a children's play area, kept in place by a wooden fence which needs to be woodstained on both sides before the sand goes in.

a The sand pit is has a radius of 2.4 m and the fence is 40 cm high. What is the area to be stained?

b The fence needs two coats of woodstain. A 2.5 litre tin claims to cover around 30 m^2. What proportion of the tin will not be used?

5 A firm sells butter in the form of a circular cylinder with radius 3.8 cm and length 12 cm. In order to wrap it fully they allow an extra 10% of the surface area for packaging. What area of packaging is used to the nearest whole number?

12 cm

3.8 cm

6 What is the surface area of this plastic seat if the inside diameter is 0.6 m and the width is 20 cm?

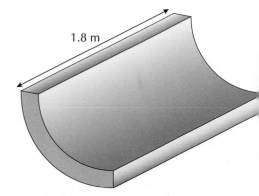

1.8 m

7 A Christmas decoration is made by cutting four quarter circles from the corners of a piece of card as shown.

The radius of each quadrant is 8 cm. Calculate the area of one side of the decoration.

Brainteaser

The diagram shows a barn with a semicircular corrugated metal roof. The lower part of the walls is made of bricks and the rest is wood panels, including the semicircle of the roof end. There is only one large wooden door, which is $\frac{1}{5}$ of its length when fully open and made completely of wood.

a How high is the highest part of the roof above the ground?
b What is the area of bricks used to build this barn?
c What is the total area of wood used, including the door?
d What proportion of the area of the whole building is the roof?

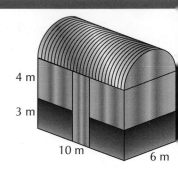

4 m

3 m

10 m

6 m

11 Solving equations graphically

11.1 Graphs from equations in the form $ay \pm bx = c$

1 Solve the equations:

  a $3x = 24$ **b** $5x = 20$ **c** $-6x = 30$ **d** $-2x = 22$

  e $2y = 12$ **f** $-9y = -27$ **g** $-5y = 45$ **h** $-4y = 8$

2 Draw the graph of each equation. Use a grid that is numbered from -10 to $+10$ on both the x-axis and the y-axis.

  a $y = x + 3$ **b** $y = x - 4$ **c** $y = 2x$ **d** $y = 3x + 1$

MR **3** **a** Draw the graphs of all these equations on the same grid. Use a grid that is numbered from -10 to $+10$ on both the x-axis and the y-axis.

    i $y = \frac{1}{2}x + 4$ **ii** $y = \frac{1}{2}x - 2$ **iii** $y = \frac{1}{2}x$

  b What do you notice about all these graphs?

  c Explain how you could now draw the graph of $y = \frac{1}{2}x + 6$.

4 Using a grid with axes numbered from -2 to 10, draw the graph of $x + y = 6$.

5 For each graph, find the coordinates of the two points where the graph intersects the x-axis and the y-axis.

  a $3x + 2y = 18$ **b** $5y - x = 15$ **c** $4y - 7x = -28$

6 Draw the graph of each equation. Use a grid that is numbered from -10 to $+10$ on both the x-axis and the y-axis.

  a $3y + 2x = 12$ **b** $4x + 5y = 40$ **c** $3y + 7x = 21$

  d $6x - y = 6$ **e** $3y - 4x = 24$ **f** $2y - 5x = -20$

7 Draw the graphs of all these equations on the same grid. Use a grid that is numbered from -10 to $+10$ on the x-axis and from -2 to $+10$ on the y-axis.

  a **i** $3y + x = 6$ **ii** $4x - 5y = -10$ **iii** $x + y = 2$ **iv** $2x - 9y = -18$

  b What do you notice about all these graphs?

8 **a** Using a grid with axes numbered from -2 to 12, draw the graph of $y = 2x + 5$.

  b Use the graph to solve these equations.

    i $2x + 5 = 4$ **ii** $2x + 5 = 8$ **iii** $2x + 5 = 10$ **iv** $2x + 5 = 2$

MR **9** **a** Complete the table for the graph $y = \frac{6}{x}$.

x	1	2	3	4	5	6
y						

  b Why can you not find the value of y when $x = 0$?

  c Using a grid with axes numbered from 0 to 6, draw the graph of $y = \frac{6}{x}$.

11.2 Graphs from quadratic equations

1 Copy and complete this table of values for $y = x^2 + 3$.

x	−1	0	1	2	3
y					

2 Copy and complete this table of values for $y = x^2 + x$.

x	−1	0	1	2	3
x^2					
x					
y					

3 **a** Copy and complete this table of values for $y = x^2 + 6x$.

x	−7	−6	−5	−4	−3	−2	−1	0	1
x^2									
$6x$									
y									

b Draw a grid with the x-axis numbered from −7 to 1 and the y-axis from −10 to 10.
c Use the table to help you draw, on the grid, the graph of $y = x^2 + 6x$.

4 **a** Copy and complete this table of values for $y = x^2 + 2x − 4$.

x	−4	−3	−2	−1	0	1	2
x^2							
$2x$							
−4							
y							

b Draw a grid with the x-axis numbered from −4 to 2 and the y-axis from −6 to 5.
c Use the table to help you draw, on the grid, the graph of $y = x^2 + 2x − 4$.

5 **a** Copy and complete this table of values for $y = 2x^2 + 1$.

x	−2	−1	0	1	2
x^2					
$2x^2$					
y					

b Draw a grid with the x-axis numbered from −2 to 2 and the y-axis from 0 to 10.
c Use the table to help you draw, on the grid, the graph of $y = 2x^2 + 1$.

6 **a** Construct a table of values for each equation. Then plot all their graphs on the same pair of axes. Number the x-axis from −2 to 2 and the y-axis from −3 to 16.
  i $y = 3x^2 - 2$ **ii** $y = 3x^2$ **iii** $y = 3x^2 + 1$ **iv** $y = 3x^2 + 3$

  b Comment on your graphs.
  c Sketch onto your diagram the graph with the equation $y = 3x^2 + 2$.

7 **a** Copy and complete this table of values for $y = 8 - 2x - x^2$.

x	−5	−4	−3	−2	−1	0	1	2	3
8									
−2x									
−x^2									
y									

  b Draw a grid with the x-axis numbered from −5 to 3 and the y-axis from −8 to 10.
  c Use the table to help you draw, on the grid, the graph of $y = 8 - 2x - x^2$.
  d Comment on the shape of the graph.

8 **a** Plot the graph of $y = \frac{(x-5)(x+5)}{2} + \frac{27}{x}$ with the x-axis numbered from 1 to 7.
  b Write down the coordinates of the minimum point on the graph of $y = \frac{(x-5)(x+5)}{2} + \frac{27}{x}$.

9 The length (L) of a pendulum is related to the period of its swing cycle (T).
  The table shows the lengths of five pendulums and their periods.

T (seconds)	2	4	6	8	10
L (metres)	1.0	4.0	8.9	15.9	24.8

  a Draw a graph with T on the x-axis and L on the y-axis.
  b Use your graph to estimate the length of a pendulum with a period of 12.2 seconds.
  c Use your graph to estimate the period of a pendulum with a length of 5 m.

Brainteaser

A circle has a circumference of 18 cm.

a Find an estimate for the radius of the circle.
b Find an estimate for the area of the circle.
c Show that, if $\pi = 3$, the area of a circle, A, is given by the formula $A = \frac{C^2}{12}$, where C is the circumference.
d Copy and complete this table of values for $A = \frac{C^2}{12}$.

C	0	6	12	18	24	30
A						

e Draw a graph with C on the x-axis and A on the y-axis.
f Use your graph to estimate the area of a circle with a circumference of:
  i 5 cm   **ii** 15 cm   **iii** 25 cm.

g Estimate the circumference of a circle with an area of 40 cm².

11.3 Solving quadratic equations by drawing graphs

1 a Copy and complete the table for the equation $y = x^2 + 2x - 1$.

x	−5	−4	−3	−2	−1	0	1	2	3
x^2									
$2x$									
-1									
y									

b Use the table to solve the equations:
 i $x^2 + 2x - 1 = 7$ **ii** $x^2 + 2x - 1 = 2$ **iii** $x^2 + 2x - 1 = -2$

2 a Copy and complete the table for the equation $y = x^2 + 3x$.

x	−5	−4	−3	−2	−1	0	1	2
x^2								
$3x$								
y								

b Use the table to solve the equations:
 i $x^2 + 3x = 0$ **ii** $x^2 + 3x = 10$ **iii** $x^2 + 3x = -2$
c Find the value of y when $x = -3.2$

3 a Draw the graph of $y = x^2 - 5x + 2$ from $x = -2$ to 7.
b Write down the value of y when $x = 4.5$
c Use the graph to find the solutions to the equations:
 i $x^2 - 5x + 2 = 0$ **ii** $x^2 - 5x + 2 = 5$ **iii** $x^2 - 5x + 2 = 10$

4 a Draw the graph of $y = x^2 + x - 3$ from $x = -4$ to 4.
b Use the graph to find the solutions to the equations:
 i $x^2 + x - 3 = -2$ **ii** $x^2 + x - 3 = 2$ **iii** $x^2 + x - 3 = 5$

5 a Draw the graph of $y = x^2 - 2x - 4$ from $x = -3$ to 5.
b Use the graph to find the solutions to the equations:
 i $x^2 - 2x - 4 = 0$ **ii** $x^2 - 2x = 1$ **iii** $x^2 - 2x = 10$

6 Draw graphs to find the solutions of:

 a $x^2 - 3x - 1 = 0$ **b** $x^2 + 4x - 6 = 2$ **c** $x^2 - x - 4 = 5$

7 Draw a graph to find the solutions of $2x^2 - 3x - 1 = 0$.

(PS) **8 a** Draw the graph of $y = 2x^2 + x - 3$ from $x = -3$ to 3.
b Write down the value of y when $x = -1.7$
c Use the graph to find the solutions to the equations:
 i $2x^2 + x - 3 = 0$ **ii** $2x^2 + x = 7$ **iii** $2x^2 + x - 2 = 0$

d Draw a straight line on your graph to solve the equations:
 i $2x^2 + x - 3 = x + 5$ **ii** $2x^2 + 2x - 7 = 0$

11.4 Solving simultaneous equations by using graphs

1 The graph shows the lines with the equations $x + y = 9$ and $y = \frac{1}{2}x + 3$.

Use the graph to solve the simultaneous equations $x + y = 9$ and $y = \frac{1}{2}x + 3$.

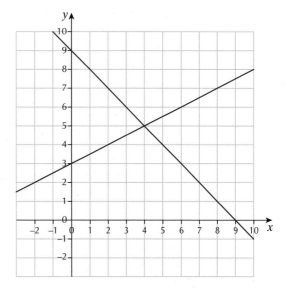

2 The graph shows the lines with the equations $3y + x = 6$ and $y = 2x + 9$.

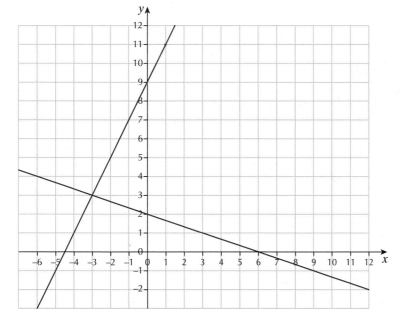

Use the graph to solve the simultaneous equations $3y + x = 6$ and $y = 2x + 9$.

3 Pair up the simultaneous equations with the answers.

a $x + y = 10$ and $2y + 3x = 23$ **i** $(-1, 6)$
b $2y - x = 20$ and $y + 4x = 1$ **ii** $(3, 7)$
c $3y + 4x = 14$ and $3y - 4x = 22$ **iii** $(8, 5)$
d $y = \frac{1}{4}x + 3$ and $y = x - 3$ **iv** $(-2, 9)$

4 Solve each pair of simultaneous equations by drawing their graphs.

a	$x + y = 4$	$y = 3x + 8$
b	$y = 2x + 5$	$y = 4x + 9$
c	$3y + x = 15$	$2y + 3x = 24$
d	$4y - 3x = 6$	$x + 2y = 8$
e	$y + 3x = 8$	$y = 3 - 1$

5 Ben and Tiara are solving the simultaneous equations $3y + 5x = 45$ and $y + 2x = 16$.

Ben says that the answer is $x = 6$ and $y = 5$.

Tiara says that the answer is $x = 3$ and $y = 10$.

Who is correct? Explain your answer.

6 a Draw the graphs of all these equations on the same grid. Use a grid that is numbered from -4 to $+10$ on both the x-axis and the y-axis.

    i $x + y = 3$ **ii** $2y = x + 3$ **iii** $y = 2x - 3$

b Use your graph to solve the simultaneous equations $x + y = 3$ and $2y = x + 3$.

c Use your graph to solve the simultaneous equations $x + y = 3$ and $y = 2x - 3$.

d Use your graph to solve the simultaneous equations $2y = x + 3$ and $y = 2x - 3$.

7 Solve the simultaneous equations

$y = x^2 - 2$ and $3x + 2y = 6$

8 Solve the simultaneous equations

$y = \frac{16}{x}$ and $2y + x = 12$

Brainteaser

The displacement (s) of a particle t seconds after being launched, at a velocity of u m/s with an acceleration of a m/s^2, is given by $s = ut + \frac{1}{2}at^2$.

Take the acceleration to be -10 m/s^2 in all questions.

a A rocket is launched from the ground vertically upwards at 60 m/s.
Find the two occasions it is 100 m above the ground.

b A beachball is launched from the ground vertically upwards at 40 m/s.
Find the time it takes to return to the ground.

c A rocket is launched from the ground vertically upwards at 37.5 m/s.
Find how long the rocket spends higher than 45 m above the ground.

d A book is dropped from the top of the Leaning Tower of Pisa.
Given that the tower is 56 m tall, find how long it takes the book to hit the ground.

e A stone is thrown vertically upwards from the top of a cliff at 10 m/s.
Given that the cliff is 75 m tall, find how long it takes the stone to splash into the water at the bottom of the cliff.

12 Compound units

12.1 Speed

1 The first humans to land on the moon travelled in the Apollo 11 spacecraft. It took approximately 73 hours to travel 385 000 km to the Moon. Calculate the average speed of the spaceship.

2 An arrow travelled 130 m at an average speed of 50 m/s. How long was the arrow in the air?

3 A train took $2\frac{1}{2}$ hours to travel from Diston to Hartley at an average speed of 50 mph. The return journey took 90 minutes.

    a Calculate the distance from Diston to Hartley.
    b Calculate the average speed over both journeys.

4 This table shows the time it takes to fly between cities.

Amsterdam				
4h 20 m	Cairo			
8h 35m	18h 40m	Chicago		
8h 15m	7h	20h 5m	Delhi	
15h 5m	10h 55m	17h 5m	6h 5m	Hong Kong

    a Delhi to Cairo is 2800 miles. What is the average speed for the journey?
    b How long does it take to fly between Delhi and Amsterdam using fractions of an hour?
    c If the average speed for this journey is 480 mph, how far apart are they?

5 Copy and complete the table which shows details of some lorry journeys.

Journey	Distance travelled	Time taken	Average speed
Marston to Surfley	80 miles	2 hours	
Deechurch to Creek	50 km		10 km/h
Penwood to Scotbridge		15 minutes	40 mph

6 Lewis and Neil ran a 2000-metre race. The distance–time graph below shows the race.

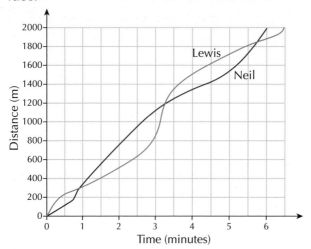

a What was Neil's average speed for the race in
   i m/s   **ii** km/h?
b After 800 m Lewis put on a spurt for about 15 seconds in order to catch Neil up. What was his speed for that part of the race in m/s and km/h?
c Use the graph to help you fill in the gaps in the report of this race:
   'Lewis started well but soon tired and Neil took the lead after about_____ seconds. Lewis made an effort to draw level at _____ metres, but could only hold the lead for _____ minutes. Neil won the race in a time of _____ minutes, and Lewis finished _____ seconds later.'

7 Some friends took part in a K1 kayak slalom competition. Over the 1 km course, there were 12 down gates and 6 up gates. It is reckoned that every down gate adds 1 second to a competitor's time and each up gate adds 3 seconds. Any missed gate adds a 15 second time penalty.

a Lucy gets through every gate successfully and finishes in 2 minutes 30 seconds. What was her average speed?
b Gemma's usual average speed without any gates is 33 km/h. Adding extra time for the gates, how long did she take if she got through every gate successfully?
c Sarah paddles at an average 36 km/h but misses one gate. In what order did these three girls finish?

8 On a journey lasting 23 minutes, this bicycle wheel rotated at a rate of 95 revolutions per minute.

a Calculate the circumference of the wheel.
b Calculate the length of the journey. Give your answer in km, correct to the nearest 100 m.

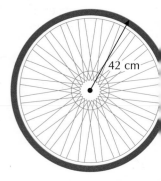

42 cm

Brainteaser

Chantal measured the distance a wind-up car travelled after different numbers of winds. This table shows her results.

Number of winds, n	5	10	15	20	30	50
Distance travelled, d metres	3.1	6.4	9.3	12.2	18.8	31.6

a   Calculate the ratio $\frac{d}{n}$ for each pair of values.
b   Is the distance travelled roughly directly proportional to the number of winds? Give a reason for your answer.
c   Write an equation connecting d and n.
d   i   Estimate how far the car would travel on 25 winds.
    ii Estimate the number of winds needed to make the car travel 24 m.

12.2 More about proportion

1   200 cm³ of butter has a mass of 172 g. Calculate the density of butter.

2   The density of oak is 0.7 g/cm³. What is the volume of a plank of oak with mass 3.2 kg?

FS 3   a   A hosepipe is accidentally left on for 15 minutes. If the water flows at 9 litres per minute, how much water is wasted?
         b   The charge for water at this house is 80p per cubic metre. How much has this wastage cost?

FS 4   a   A Volkswagen takes 35 seconds to fill its fuel tank with 70 litres of diesel. What is the rate of flow of the fuel at this pump?
         b   At the same pump, a Ford Focus takes 22 seconds to fill up. How much fuel does it hold?
         c   A Toyota holds 16 gallons of petrol. How long does it take to fill up, to the nearest second?
         d   If diesel costs £1.39 per litre and petrol costs £1.32, which of the three cars costs most to fill up?

5   A hosepipe fills this cylindrical can at a rate of 15 litres/minute.

a   Calculate:
    i   the volume of the can in cubic centimetres
    ii the capacity of the can in litres.
b   How long does it take to fill the can? Give your answer to the nearest second.

20 cm

30 cm

6 This graph shows the connection between mass and volume for a type of steel.

    a Use the graph to find the mass of 1.25 m³ of steel.
    b Use the graph to find the volume of 3000 kg of steel.
    c What is the density of this steel?

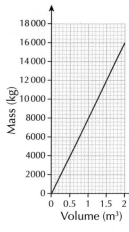

7 An alloy is made from 3.2 kg of copper, 1.5 kg of lead and 600 g of tin. Calculate the density of the alloy.

Metal	Density (g/cm³)
copper	8.96
lead	11.4
tin	7.3

Brainteaser

The table shows some facts about several of the planets in our solar system. Copy it and fill in the blanks. Note the different units in the density columns to guide you.

Planet	Volume (km³)	Mass (kg)	Density (kg/km³)	Density (kg/m³)	Density (g/cm³)
Earth	1.1×10^{12}	6×10^{24}	5.455×10^{12}	5455	
Jupiter	1.4×10^{27}				1.326
Saturn	8.3×10^{14}	5.7×10^{26}		687	

12.3 Unit costs

1 1 kg of margarine costs £2.50.

    a Calculate the cost per 100 g.
    b Calculate the cost per gram.
    c Calculate the number of grams bought for 1p.
    d Calculate the number of grams bought for £1.

2 Six boxes contain 54 light bulbs. How many light bulbs are there in:

    a one box   **b** nine boxes?

3 Jason used 6 litres of paint to cover 15 m of fence.
How much paint is needed to cover 40 m of fence?

FS **4** Which is better value for money: 2 litres of milk at £1.45 or 4 pints at £1.59?

5 Miriam downloaded 928 kB of data from the Internet in 18 seconds.

  a How much data did she download each second, to the nearest kB?
  b How much data could she download at the same rate in:
    i 45 seconds   **ii** 7 minutes?

FS **6** **a** Eighteen eggs cost a shopkeeper £3.60. How much does each egg cost him?
  b He puts them in boxes of six and charges £1.80 per box. How much profit does
PS he make on each box? Express this as a percentage of the cost.
  c He does an offer of three boxes for five pounds. What is his percentage profit
  here?

MR **7** The diagram shows how two businesses
charge for the hire of a canoe.

FS
  a For which business is the charge
  directly proportional to the hire
  period?
  Give a reason for your answer.

  b For this business, what is the hire
  charge for one hour?

  c For this business write an equation
  connecting the hire charge, H, with
  the hire period, p.

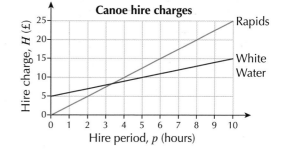

  d **i** What would this company charge to hire the canoe for 4.5 hours?
    ii What is the hire period corresponding to a hire charge of £18 with this
    company?
    Give your answer in hours and minutes.

Brainteaser

These tables show the amount charged for different weights of cheddar cheese in two shops.

Country Fare					
Weight, W grams	200	350	500	600	900
Cost, C pence	84	149	210	243	369

Farm Fresh					
Weight, W grams	150	400	550	700	1200
Cost, C pence	66	176	242	308	528

a   Calculate the ratio $\frac{cost}{weight}$ for each pair of values.

b   For which shop is the cost directly proportional to the weight of cheese?
    Give a reason for your answer.

c   Write an equation connecting C and W for this shop.

d   i   How much would this shop charge for 1.6 kg of cheese?
    ii   What weight of cheese can be bought for £5 at this shop?

13 Right-angled triangles

13.1 Introducing trigonometric ratios

1 a Draw three different sized right-angled triangles, but keep the angle at the bottom right the same in each triangle, as shown.

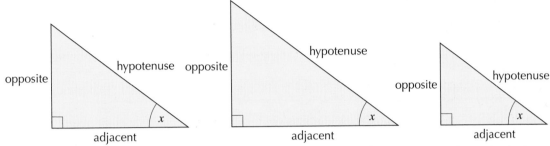

Hint: If you choose the lengths to be a whole number of centimetres, the calculations are simpler.

b Label your triangles A to C.
c Copy the table below.

Triangle	Opposite	Adjacent	Hypotenuse	Opposite Hypotenuse	Adjacent Hypotenuse	Opposite Adjacent
A						
B						
C						

d Measure the sides for each triangle and fill in their lengths on the table.
e Work out the answers to the three end columns as decimals.
f What do you notice about the results in each of the three end columns?

2 Draw two right-angled triangles that both have the other angles equal to 60° and 30°.

a Find the sine and cosine of both these angles.
b What do you notice?
c Why is this?

3 Here are some other trigonometric ratios. What do you notice in each set?

a sin 0°, sin 25°, sin 45°, sin 70°, sin 90°
b cos 90°, cos 65°, cos 45°, cos 20°, cos 0°
c Compare your answers to parts **a** and **b** above. What can you say about the relationship between sines and cosines in the light of these results?

4 Make some observations about these trigonometric ratio sequences.

a tan 0°, tan 15°, tan 30°, tan 45°, tan 60°, tan 75°
b tan 88°, tan 89°, tan 89.5°, tan 89.8°, tan 89.9°, tan 90°
c What happens to tangent ratios that does not happen with sine or cosine? Can you explain this?

5 Multiply these trigonometric expressions by the number shown to find their values, where possible, correct to three decimal places.

    a 200 sin 90°   **b** 200 sin 0°   **c** 200 sin 30°

    d 500 cos 90°   **e** 500 cos 0°   **f** 500 cos 60°

    g 350 tan 45°   **h** 350 tan 0°   **i** 350 tan 90°

13.2 How to find trigonometric ratios

1 Copy these triangles and labels the sides with the words opposite, adjacent and hypotenuse.

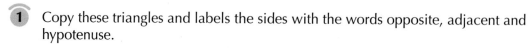

2 For each triangle, identify the opposite, the adjacent and the hypotenuse in relation to the angle labelled θ. (For example, d = hypotenuse)

a
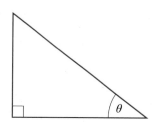
b
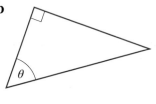
c
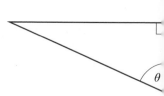

(MR) **3** For each part, draw a right-angled triangle with the correct sides labelled from the information given:

    a $\sin \theta = \frac{3}{5}$ **b** $\cos \theta = \frac{12}{13}$ **c** $\tan \theta = \frac{9}{12}$ **d** $\tan \theta = \frac{12}{9}$

4 For each triangle, complete the trigonometric ratios, leaving your answer as a decimal to two decimal places. The first one has been done for you.

a

b
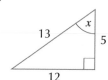
c
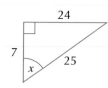

    a $\sin x = \frac{\text{opp}}{\text{hyp}} = \frac{8}{10} = 0.8$

Now find cos and tan for part **a**, then do all three for parts **b** and **c**.

13.3 Using trigonometric ratios to find angles

1. Use inverse sine or cosine to find angle θ, giving your answers correct to one decimal place.

   a $\sin \theta = 0.292$
   b $\sin \theta = 0.588$
   c $\sin \theta = \frac{5}{8}$
   d $\sin \theta = \frac{8}{9}$

   e $\cos \theta = 0.139$
   f $\cos \theta = 0.875$
   g $\cos \theta = \frac{7}{15}$
   h $\cos \theta = \frac{5}{16}$

2. Find angle θ, giving your answers correct to one decimal place.

   a $\tan \theta = 0.577$
   b $\tan \theta = 1.732$
   c $\tan \theta = 4.5$
   d $\tan \theta = 18.6$

   e $\tan \theta = \frac{2}{3}$
   f $\tan \theta = 1\frac{3}{4}$
   g $\tan \theta = 17\frac{1}{2}$
   h $\tan \theta = \frac{24}{18}$

3. These questions all use the sines of angles. Use your calculator to find each angle, correct to one decimal place.

   a 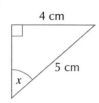 **b** 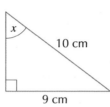 **c** 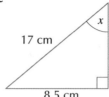 **d**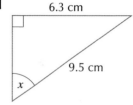

4. Use your calculator and cosines to find each angle, correct to one decimal place.

   a 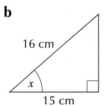 **b** 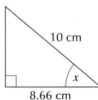 **c**   **d**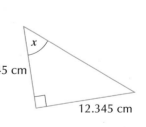

5. Use tangents to find each angle, correct to one decimal place.

   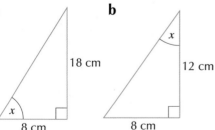 

   (Hint: Be careful to get the numerator and denominator the right way round.)

6 Calculate the size of the angle labelled x in each triangle. Give your answers correct to one decimal place. You will need to decide whether to use sin, cos or tan each time.

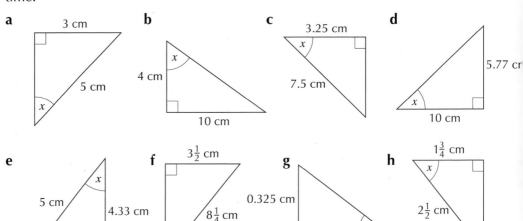

a 3 cm, 5 cm, x

b 4 cm, 10 cm, x

c 3.25 cm, 7.5 cm, x

d 5.77 cm, 10 cm, x

e 5 cm, 4.33 cm, x

f $3\frac{1}{2}$ cm, $8\frac{1}{4}$ cm, x

g 0.325 cm, 0.875 cm, x

h $1\frac{3}{4}$ cm, $2\frac{1}{2}$ cm, x

Draw diagrams of your own to help with the next few questions.

7 **a** In the triangle ABC, A = 90°, AB = 20 mm and AC = 54 mm.
Find angle B. Give your answer correct to one decimal place.

b In the triangle PQR, P = 90°, PR = 9 cm and QR = 24 cm.
Find angle R. Give your answer correct to one decimal place.

(PS) **8** A train travelled 2 km along a straight track and climbed 300 m vertically.
Calculate the angle the track makes with the horizontal.

Brainteaser

A dinghy is somewhere on a line 750 m south of a ship, but is 1200 m away in a straight line.
Calculate the angle at the boat's position between the dinghy and due south.

13.4 Using trigonometric ratios to find lengths

1 Find the values of these trigonometric expressions, giving your answers correct to three decimal places.

a sin 16°   **b** sin 69°   **c** sin 13.4°   **d** cos 74°
e cos 28°   **f** cos 2.5°   **g** tan 1°   **h** tan 27°
i tan 45°   **j** tan 89.3°

2 Calculate the following, giving your answers correct to three significant figures.

    a 10 sin 67 ° **b** 8.5 sin 40   **c** 4 sin 28°

    d 5 cos 50°   **e** 6 cos 22.5° **f** 3.5 cos 88.3°

    g 7 tan 12°   **h** 1.5 tan 38.5° **i** 100 tan 87°

3 Calculate x correct to three significant figures.

    a   **b**   **c**

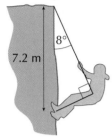
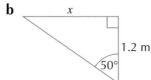
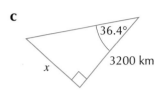

4 The diagram shows an abseiler. Calculate the length of the rope.

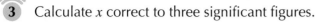

5 Sanchez drew the line AB on a computer using a graphing program.

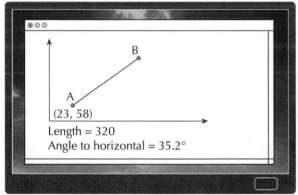

    The program shows the length and angle of the line. Calculate the coordinates of point B.

Draw diagrams of your own to help with the next few questions.

6 **a** In the triangle LMN, L = 33°, M = 90° and LM = 1.7 m.

        Find the length of MN. Give your answer correct to three significant figures.

    b In the triangle XYZ, X = 90°, Y = 68° and YZ = 42 mm.

        Find the length of XZ. Give your answer correct to three significant figures.

7 A rambler walked 6 km from Odale to Hype at an angle of 23° from north.

    How far east is Hype from Odale?

8 A model aeroplane is attached to the top of a 3 metre pole by a 2 metre string and flies in a circle 1.6 metres above the ground.

    a Calculate the angle the string makes with the pole.

    b Calculate the distance between the aeroplane and the pole.

  9 An equilateral triangle has a side of 12 cm. Use trigonometry to calculate the shortest distance from a vertex to its opposite side.

Brainteaser

The diagram shows a regular octagon.

    a Use your knowledge of interior or centre angles to work out the distance between opposite corners.

    b Calculate the distance between opposite sides.

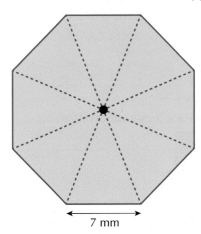

7 mm

14 GCSE preparation

14.1 Number

1 Shania works 28 hours a week for £7.31 per hour. Sui Main works 43 hours a week for £4.62 per hour.

    a How much does each person earn per week? Work in pence.
    b What is the difference in their weekly earnings? Work in pence.

2 Hotel rooms cost £23 per person per night.

    a How much will it cost for one person to stay for 14 nights?
    b How much will it cost for three people to stay for 9 nights?
    c Samuel's hotel bill was £552. How many nights did he stay?
    d The hotel was fully booked for three days and took £828.
       How many people stayed at the hotel each night?

3 Jan threw a dart at a dartboard 84 times. $\frac{3}{7}$ of the darts hit the inner ring, $\frac{1}{6}$ hit the outer ring, $\frac{3}{14}$ hit the bullseye, and the remainder hit doubles or trebles.

    a How many darts hit
      i the inner ring   **ii** the outer ring
      iii the bullseye   **iv** a double or treble?
    b What fraction hit a double or treble?

4 $\frac{1}{6}$ of a magazine is used for adverts, $\frac{2}{9}$ is devoted to a feature, and the rest is used for regular articles.

    a What fraction is used for regular articles?
    b The magazine has 54 pages. How many are devoted to the feature?

5 A reel contains $3\frac{3}{8}$ metres of flex. Jane buys $1\frac{5}{8}$ metres. What fraction of the reel did she buy?

6 Debbie measured these pipes using an old ruler.

 $2\frac{9}{10}$ inches

 $2\frac{3}{16}$ inches

 $2\frac{7}{12}$ inches

    a What is the total length of pipes A and B?
    b How much longer is pipe C than pipe B?
    c Which two pipes have the closest length?

7 A swimming pool contains 13 450 litres of water.

   a A pump increases the volume in the pool by 7%.
      How much water does it now contain?

   b 13% of the water is then drained away.
      How much water does it now contain?

   c Every week, the pool loses 1.4% of its water due to evaporation.
      How much water does it contain
        i after one week   **ii** after four weeks?

(FS) **8** **a** What is the cost of the jacket including VAT?

   b What was the cost of the trousers before VAT was added?

£94 excluding VAT @ 20%

£58 including VAT @ 20%

9 The speed of light is approximately 3×10^8 m/s.

How long does it take for light to travel one metre?

Give your answer in standard form, correct to 3 significant figures.

10 Find the answer to each sum, giving the answer in standard form.

   a $(3 \times 10^9) + (6 \times 10^7)$ **b** $(3 \times 10^9) - (6 \times 10^7)$
   c $(3 \times 10^9) \times (6 \times 10^7)$ **d** $(3 \times 10^9) \div (6 \times 10^7)$

14.2 Algebra

1 Simplify these expressions.

   a $3x - 5 + 6x + 6$ **b** $a + 4 - 3a - 7$ **c** $3y - -2y$
   d $3d + 4e - 3e + 2d$ **e** $2 \times (-4t)$ **f** $5x \times 4x$
   g $3ab - 2a + 4ab - 5a$ **h** $4y^2 - 2y + 6y^2 + 5y$

2 Expand and simplify these expressions.

   a $3(2x - 1)$ **b** $-(4y + 7)$ **c** $x(3 + 2x)$ **d** $-4(2 - 5y)$
   e $4 + 2(3a + 2)$ **f** $6x - 3(x - 5)$ **g** $2d(e + 3) - d$ **h** $5p^2 - 3p(p - q)$

3 Solve these equations.

   a $5t - 2 = 13$ **b** $\frac{x}{4} + 3 = 9$ **c** $7d = 4d + 21$ **d** $2m - 4 = m - 6$
   e $-8w = 7$ **f** $2p = 9p - 35$ **g** $3(x - 3) = 18$ **h** $-2(3y + 4) = 16$

4 Solve these equations.

   a $2(x + 3) + 3(x - 1) = 23$ **b** $2(2t - 1) + 3(t + 8) = 1$
   c $4(2y + 3) - 2(3y - 1) = 18$ **d** $2(p + 3) - 4(p - 1) = 7$
   e $9(m + 2) = 2(3m + 8) + 11$

5 **a** Write an equation about this rectangle involving x.
        Solve your equation.

   b **i** Write an equation about this pie chart involving x.
           Solve your equation.
      ii Find the percentage of people who chose football
           as their favourite sport.

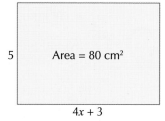

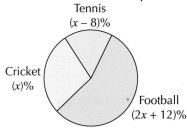

6 **a** Draw the graphs of all these equations on the same grid. Use a grid that is
      numbered from −4 to 10 on the x-axis and 0 to 10 on the y-axis.

     i $y = \frac{1}{2}x + 3$ **ii** $x + y = 9$ **iii** $y = 2$

   b Use your graphs to solve each pair of simultaneous equations.

     i $y = \frac{1}{2}x + 3$ $x + y = 9$
     ii $x + y = 9$ $y = 2$
     iii $y = 2$ $y = \frac{1}{2}x + 3$

7 Factorise each expression.

   a $7m - 14$ **b** $12m^2 + 16m$ **c** $15m + 5n$ **d** $20mn - 14m$

8 **a** Find the equation of the line parallel to $y = 3x - 4$ that passes through the point
      (6, 11).

   b Find the equation of the line perpendicular to $y = 3x - 4$ that passes through the
      point (6, 11).

9 Find the area of the rectangle.

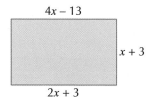

10 Solve these equations.

   a $x^2 + 12x + 35 = 0$ **b** $x^2 + 9x + 20 = 0$ **c** $x^2 + 9x + 18 = 0$
   d $x^2 + 8x - 33 = 0$ **e** $x^2 - 14x + 13 = 0$ **f** $x^2 + 4x - 32 = 0$
   g $x^2 - 14x + 48 = 0$ **h** $x^2 - 12x + 36 = 0$

14.3 Ratio, proportion and rates of change

1 Warren types 54 words per minute. How many words does he type in

    a 19 minutes   **b** 48 minutes   **c** 2 hours   **d** 4 hours 17 minutes?

2 Which purchase is the best value? Work in pence.

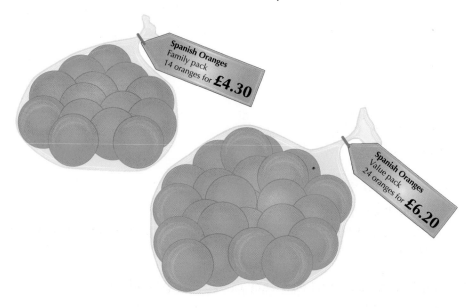

3 Quentin used 629 cl of paint to colour 37 m² of floor. Each metre of floor is made from 16 square tiles. How much paint is needed to cover:

    a one square metre of floor   **b** a floor of area 930 m²   **c** 448 tiles?

4 A cupcake recipe uses the following ingredients.

Makes 12 cupcakes

180 g butter

240 g caster sugar

4 eggs

200 g flour

    a How much butter would be required to make 48 cupcakes?
    b How much caster sugar would be required to make 30 cupcakes?
    c Write down the ingredients required to make 9 cupcakes.

5 A mango cordial is made using a 2 : 7 ratio of concentrate : water.

Vanessa wants to make some mango cordial.

    a How much concentrate would be required if she has 560 ml of water?
    b How much water would be required if she has 50 ml of concentrate?

Vanessa is hosting a tea party and wants to make 36 jugs of cordial.

    c How much concentrate does she need?

6 The graph below shows how to convert between kilograms and pounds.

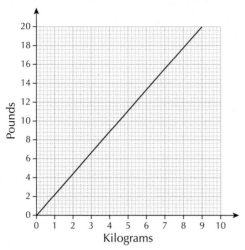

a Use the graph to convert:
   i   5 kilograms into pounds
   ii 14 pounds into kilograms.

b Deepak weighs 75 kg.

Given that there are 14 pounds in 1 stone, find an estimate for Deepak's mass in stones and pounds.

  7 Tharushi's car has a fuel efficiency of 42 miles per gallon.

1 mile = 1.609 km

4.55 litres = 1 gallon

At Tharushi's local petrol station, petrol costs £1.33 per litre.

How many kilometres can Tharushi drive for £50?

8 A coach leaves the depot at 10.20 am and sets off for a town 90 miles away at an average speed of 40 mph. Four and a half hours later, the coach returns to the depot at an average speed of 50 mph.

a How long does the outward journey take?
b At what time does the coach return to the depot?
c Illustrate the information in a distance–time graph.

14.4 Geometry and measures

1 State the number of lines of symmetry and the order of rotational symmetry for each letter.

A N H R S D

(MR) **2** ABC is a triangle.

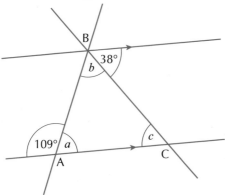

a Find the size of angle a. Give a reason for your answer.
b Find the size of angle c. Give a reason for your answer.
c Explain how you know that triangle ABC is isosceles.

3 A rectangle has a base of 14 cm and a height of 9 cm.

A triangle has a base of 7 cm.

The rectangle and the triangle have the same area.

Find the perpendicular height of the triangle.

(MR) **4** Shayla has a circular table which has a diameter of 1.7 m.

a Find the circumference of the table, correct to one decimal place.

Shayla has bought a circular tablecloth which has an area of 2.4 m².

b Is the tablecloth large enough to fit the table? Explain your answer.

5 The cross-section of a prism is a quadrant (quarter of a circle) with a radius of 7 cm.

The length of the prism is 13 cm.

a Find the volume of the prism, correct to the nearest integer.

The prism is made from steel. The density of steel is 7.6 g/cm³.

b Find the mass of the prism, correct to the nearest integer.

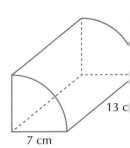

13 c

7 cm

6   **a**   Enlarge the trapezium WXYZ using A as the centre of enlargement and a scale
factor of 3. Label the image W′X′Y′Z′.

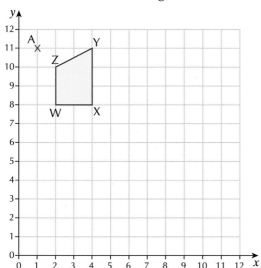

b   Write down the coordinates of the vertices of the image.
c   The area of WXYZ is 50 cm². Find the area of W′X′Y′Z′.

7   ACE and BCD are triangles.

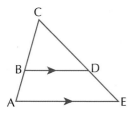

a   **i**   Explain why angle EAC is the same as angle DBC.
 ii   Explain how you can tell that triangles ACE and BCD are similar triangles.

AB = 5 cm, BC = 10 cm, CD = 12 cm and BD = 14 cm.

b   Find the length of CE.
c   Find the length of AE.

8   A square is surrounded by four identical regular polygons.

How many sides does the polygon have? Explain your answer.

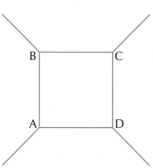

9 **a** PQR is a right-angled triangle.

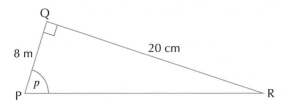

     i Find the length of PR, correct to three significant figures.
     ii Find the size of angle p, correct to three significant figures.

  b STU is a right-angled triangle.

     i Find the length of ST, correct to three significant figures.
     ii Find the size of angle u, correct to three significant figures.

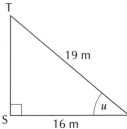

10 The diagram shows a sector with an angle of 72° and a radius of 15 m.

  a Find the area of the sector, giving the answer in terms of π.
  b Find the perimeter of the sector, giving the answer in terms of π.

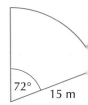

14.5 Probability

1 Put these events in order of likelihood from least likely to most likely.

  A Rolling a 5 on a fair six-faced dice numbered 1 to 6

  B Rolling a number less than 10 on a fair six-faced dice numbered 1 to 6

  C Rolling a number greater than 1 on a fair six-faced dice numbered 1 to 6

  D Rolling a 7 on a fair six-faced dice numbered 1 to 6

  E Getting heads when tossing a fair coin

2 A letter is chosen at random from the word HIPPOPOTAMUS.

Find the probability that the letter is

  a a P   **b** an O   **c** a G   **d** a vowel   **e** not an M.

3 Jennie will have either toast, cereal, fruit or a croissant for her breakfast.

The table shows the probabilities of what Jennie will choose for her breakfast.

Breakfast	toast	cereal	fruit	croissant
Probability	0.3	0.45	0.2	

  a Find the probability that Jennie will have a croissant for her breakfast.
  b Find the probability that Jennie will not have cereal for her breakfast.
  c Find the probability that Jennie will have either cereal or fruit for her breakfast.
  d Estimate the number of days in April that Jennie had toast for her breakfast.

4 Bruce throws two five-sided dice, each numbered 1, 2, 3, 4, 5.

    a Draw a table to show all the possible results, if the numbers on each of the dice are added together.

    b When the two dice are thrown, what is the probability that:
       i both dice show the same number
      ii the total score is an odd number
     iii the total score is at least 4
     iv the total score is not a prime number?

 5 Six friends investigated the probability of rolling a 6 on a biased die.

Their results are recorded as follows.

Person	Total number of rolls	Number of 6s
Ted	54	15
Charlotte	42	14
Anais	75	23
Lois	8	2
Emily	25	9
Albert	18	4

    a Use Lois' data to find an estimate for the probability of rolling a 6.

    b Whose data is the most reliable for finding an estimate for the probability of rolling a 6? Explain your answer.

    c Use all the data to find an estimate for the probability of rolling a 6.

 6 State whether the outcomes in each pair are mutually exclusive or not mutually exclusive.

Explain your answers.

    a An ordinary, six-sided dice landing on an odd number.
       An ordinary, six-sided dice landing on an even number.

    b An ordinary, six-sided dice landing on a square number.
       An ordinary, six-sided dice landing on an even number.

    c An ordinary, six-sided dice landing on a prime number.
       An ordinary, six-sided dice landing on an even number.

 7 The probability of getting 8 on a spinner is 40%.

Aaron spins the spinner 300 times.

Estimate the number of times he should expect to get an 8.

8 A bag contains red, green and blue beads in the ratio 1 : 2 : 3.
Allan takes a bead out of the bag at random.

    a What is the probability that the bead is green?

Michael counts the remaining beads and then puts them back in the bag.
There are 59.

Michael takes a bead out of the bag at random.

    b If Allan's bead was blue, what is the probability that Michael's bead is red?

    c If Allan's bead was blue, what is the probability that Michael's bead is blue?

9 A shelf has two fiction books and three non-fiction books.

Toby takes two books off the shelf.

What is the probability that

a both books are fiction
b both books are non-fiction
c one book is fiction and the other is non-fiction?

14.6 Statistics

 1 Joseph has collected data on his class' favourite colours.

Colour	Frequency
blue	5
green	11
purple	2
orange	9
red	3

Represent the information in a pie chart.

2 Six children took a Geography test. Five of them scored 17, 13, 16, 12 and 19.

Their mean score was 15.

a Find the missing score.   b What was the median score?
c What was the modal score?   d What was the range of the scores?

 3 The scatter diagram shows the number of hours of TV watched by several students and their scores in a Mathematics test.

A line of best fit has been drawn on the scatter diagram.

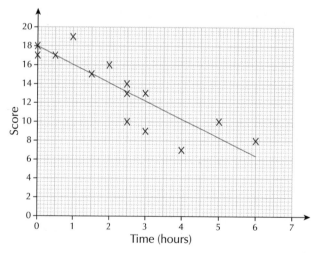

a Describe the correlation between time spent watching television and test score.
b Anna watched television for 3 hours 30 minutes. Estimate her test score.
c John scored 12 on the test. Estimate how long he spent watching television.
d Ryan says that he watched 12 hours of television. Explain why it would not be appropriate to use the scatter diagram to estimate his score.

4 Norman wanted to find out how often people go to the cinema.

He wrote this question in a questionnaire.

How many times have you been to the cinema?

1 – 4 ☐   4 – 7 ☐   7 – 10 ☐

   a State three things that are wrong with this question.
   b Write a question that corrects these mistakes.

5 Here are seven algebraic expressions.

$$x + 5, \ x - 3, \ x + 4, \ x + 11, \ x - 9, \ x - 3, \ x + 9$$

   a Find the range.
   b What is the median?
   c What is the mode?
   d Find the mean.

6 State five numbers which obey all these properties:

- the mean is 36
- the range is 7
- the mode is 39
- the median is 37

7 The table shows how many goals were scored by a football team in 38 matches.

Number of matches	14	10	5	6	3
Number of goals	0	1	2	3	4

   a Find the total number of goals scored.
   b What is the modal number of goals scored?
   c Find the mean number of goals scored per match.
   d Find the median number of goals scored.

8 Madeleine is researching the prices of computers on the Internet.

She records her findings in a table.

Price (£)	Frequency
$100 \leqslant P < 300$	6
$300 \leqslant P < 500$	10
$500 \leqslant P < 700$	3
$700 \leqslant P < 900$	1

   a State the modal class.
   b In which class is the median price?
   c Find an estimate for the mean price.

9 The masses of 80 turnips grown by a farmer are recorded in the table.

Mass (g)	Frequency	Cumulative frequency
$80 \leqslant m < 90$	4	
$90 \leqslant m < 100$	10	
$100 \leqslant m < 110$	35	
$110 \leqslant m < 120$	22	
$120 \leqslant m < 130$	7	
$130 \leqslant m < 140$	2	

a   Copy and complete the cumulative frequency table.
b   Represent the data in a cumulative frequency graph.
c   Use your graph to find an estimate for the median.
d   Use your graph to find an estimate for the interquartile range.

The heaviest 5% can be entered for an oversize vegetable competition.

e   Find the lightest mass a turnip can be to be entered.

Mixed GCSE-style questions

1 At The Eden Project, the biomes cover a total of 22130 m² and contain about 135 000 plants.

   a Find the planting density (the number of plants per square metre) of the biomes.

   b Is it reasonable to think that every square metre has this planting density? Explain your answer.

2 **a** Using a ruler and a protractor, make an accurate copy of this diagram of a regular hexagon.

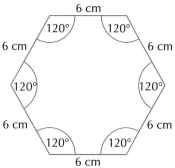

   b The diagram below shows a hexagon that is 11 m across. The length of each side is 5.5 m.

     A hexagon can be made from two identical trapeziums.

     The formula for the area of a trapezium is:

     $\frac{1}{2}(a + b)h$

     Find the area of the hexagon above.

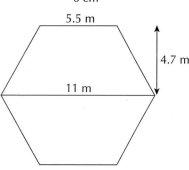

3 The graph below shows the change in temperature for part of one day last June in the Humid Tropics Biome (HTB) at The Eden Project.

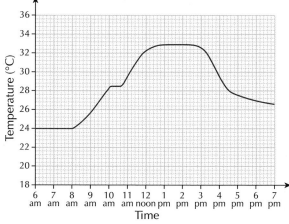

   a What time did the Sun rise?

   b The Sun went behind a cloud in the morning. For how long?

   c The automatic vents start working when the temperature rises too high. At what time did they start working?

   d What was the minimum temperature in the HTB?

   e What was the maximum temperature in the HTB?

   f What do you think happens to the temperature in the HTB after 7 pm?

4 Here are some road signs.
For each part, copy the sign and draw in all the lines of symmetry.

a  b  c

5 Here are some road signs.
For each part, state the order of rational symmetry.

a 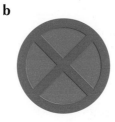 b 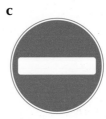 c

6 The table shows the percentage of casualties in road accidents for different age groups.

Age	Percentage
17–25 years	33%
26–39 years	28%
40–59 years	24%
60 years and over	15%

a Draw a pie chart to show this information.
b Give a reason why there are more casualties in the 17–25 years age group.

7 A driving test lasts for 40 minutes and covers 12 miles. Work out the average speed miles per hour.

8 Here is a formula for working out stopping distance, d (feet), when travelling at speed v (mph).

$$d = v + \frac{v^2}{20}$$

a Work out the stopping distance in feet when $v = 20$ mph.
b By working out the stopping distance in feet when $v = 30$ mph, show that a car would travel almost twice as far when stopping from 30 mph than 20 mph.

9 A study on the body weights of squirrels gave the following data for red squirrels over a 12-month period.

Month	Jan	Feb	Mar	Apr	May	Jun	Jul	Aug	Sep	Oct	Nov	Dec
Average weight (g)	273	265	274	280	285	290	310	325	345	376	330	290

This graph shows the same data for grey squirrels.

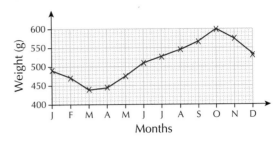

a On a copy of the graph, draw a line graph to show the average body weight of the red squirrels.

b Why do you think the weights of the squirrels increase in the autumn?

c Comment on the differences in the weights of the red and grey squirrels over the year.

10 The scatter diagram shows the relationship between the body length and tail length of red squirrels.

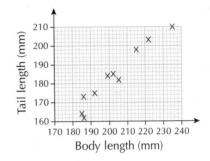

a Estimate the tail length of a red squirrel with a body length of 210 mm.

b Describe the correlation between the body length and tail length of red squirrels.

Grey squirrels

	A	B	C	D	E	F	G	H	I	J
Body length (mm)	272	243	278	266	269	280	251	272	278	281
Tail length (mm)	223	196	220	218	218	222	198	220	226	225

c i Using the data for grey squirrels, draw a scatter diagram to show the relationship between body length and tail length.
Use a horizontal axis for body length from 240 mm to 290 mm and a vertical axis for tail length from 190 mm to 230 mm.

ii Draw a line of best fit on the diagram.

iii Use your line of best fit to estimate the body length of a grey squirrel with a tail length of 205 mm.

iv Explain why the diagram could not be used to estimate the tail length of a young grey squirrel with a body length of 180 mm.

11 This table shows the number of baby red squirrels born 100 nests.

Number of baby squirrels	1	2	3	4	5
Frequency	24	42	19	10	5

a Work out the mean number of baby squirrels per nest.

b What is the median number of baby squirrels per nest?

c The probability of a baby squirrel surviving to adulthood is 0.4.
How many of the squirrels in the table above would you expect to survive to adulthood?

12 A male red squirrel can roam over an area of up to 42 acres. This is about the size of 34 football pitches. Assuming that a football pitch is 100 m by 50 m, calculate (to 2 decimal places) how many acres are in one hectare (10000 m²).

13 Jeff and Donna are planning to set up a mobile shop. They have bought a bus to run the shop from.

They know that fitting out the bus will cost £5000.

Insuring and taxing the bus will cost £1500.

Buying the initial stock for the shop will be £2000.

They have savings of £7500.

   a What is the minimum amount they need to borrow from the bank to get started?

   b The bank agrees to lend them up to £10 000 at an interest rate of 7.5% per annum.

     i How much will the interest be on £10 000 for one year?

     ii If they only borrow the minimum amount they need and decide to pay the loan back over one year, what will be the approximate monthly payment?

14 Jeff and Donna's bus is 12 m long and 2.5 m wide.

This is a scale drawing of the bus with 1 cm representing 1 m.

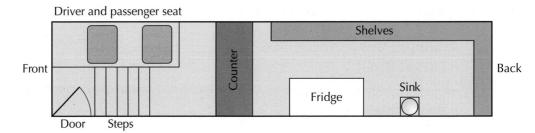

   a What is the total area of the bus?

   b What is the actual width of the counter?

   c The area behind the counter is the 'shop area'. What percentage of the total area is the 'shop area'?

     Give your answer to the nearest percentage.

   d The fridge is 1 m high. What is the volume of the fridge?

15 This table shows how much households in a village spend each week on meat.

Amount, m, £	Frequency, f	Midpoint, m	m × f
£0 ≤ m < £10	48	5	
£10 ≤ m < £20	97	15	
£20 ≤ m < £30	38	25	
£30 ≤ m < £40	17	35	
Total		Total	

Copy the table and work out the mean amount each household spends on meat each week.

16 These are the prices of seats at a theatre.

	Full price	Concessions (children and senior citizens)
Monday to Thursday	£16.50	£11.75
Friday and Saturday	£22.50	£16.75

A family is going to the theatre.

There are:

- two adults, who will pay full price
- a grandmother, who is a senior citizen
- three children.

They are going on Wednesday.

Calculate the price of the tickets.

17 The bar chart shows the numbers of students absent each day of the week. They are all in the same year group.

a Find the mean number of students absent each day.

b There are 160 students in the year group. The target for the school is to have less than 7% absent on each day. On how many days did the school meet its target?
Give a reason for your answer.

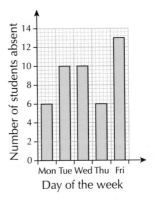

18 This pie chart shows the grades achieved by all the candidates in one school in a GCSE examination.

What proportion of those who gained A*-C were awarded an A*?

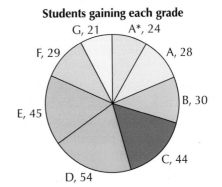

Students gaining each grade

19 These are the prices of some items in a sale.

Item	Original price	Sale price
Formal shirt	£70.00	£22.95
Casual shirt	£80.00	£22.95
Polo shirt	£56.00	£22.95
Handmade tie	£90.00	£29.50

Harry buys a formal shirt, a casual shirt and two ties.

Calculate his percentage saving, based on the original prices.

20 The cost of electricity is in two parts. Each unit used costs 13.38p. There is a standing charge of 14p per day. The number of units used is the difference between two meter readings.

At the start of a 93-day period the meter reading is 25 643.

At the end of that period the meter reading is 26 178.

Calculate the cost of the electricity used in that 93-day period.

21 The graph shows the stopping distances, in metres, for a car driving on a dry, level road at different speeds.

Car A and Car B are travelling on a motorway. Car A is travelling at 70 mph.

The stopping distance for Car B is two-thirds of the stopping distance for Car A. Find the speed of Car B.

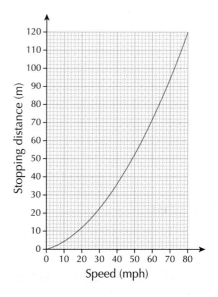

22 On a farm there are goats, sheep and chickens.

The ratio of goats to sheep is 2 : 3.

The ratio of sheep to chickens is 2 : 5.

What is the ratio of goats to chickens?

23 Here are the first four patterns in a sequence.

There are two squares in pattern number 1 and six squares in pattern number 2. Explain how you could work out the number of squares in any pattern.

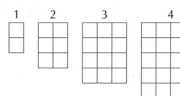

24 This shape is cut from a square of card. All the angles are right angles.

Find an expression for the perimeter of the shape, in terms of a and b.

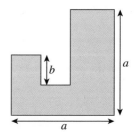

25 You can buy euros **from** a bank or sell them **to** the bank.

If you buy them from the bank you can get €1.09 for £1.00.

If you sell them to the bank you can get £1.00 for €1.23.

How much will you lose if you buy €100 from the bank and then sell them back?

Give your answer in pounds.

26 Jasmine is thinking of a two-digit number.

What are the possible values of Jasmine's number?

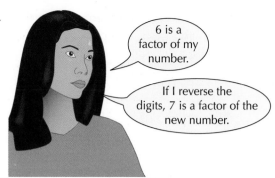

> 6 is a factor of my number.

> If I reverse the digits, 7 is a factor of the new number.

27 Arthur is using loft boards to put a floor in his loft. Loft boards are rectangular panels 69 cm long and 32 cm wide.

The floor is rectangular and measures 3 m by 4 m.

Arthur can cut the boards if necessary.

He has 50 loft boards. Is this enough?

Give a reason for your answer.

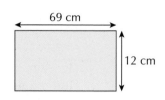

69 cm

12 cm

28 All the employees in a firm are given the same percentage pay rise.

Alan's monthly salary increases from £2450.00 to £2508.80.

Betty's monthly salary before the increase is £3055.00.

What is Betty's monthly salary after the pay rise?

29 Two cars are on a motorway travelling at 100 km per hour (63 miles per hour).

The cars are 40 metres apart.

Use the two-second rule to decide whether this is a safe distance.

Give a reason for your answer.

Attention all drivers!
STAY SAFE!
Follow the two-second rule
Stay at least 2 seconds
behind the car in front of you

40 m

30 Here are two cylindrical cans of food.

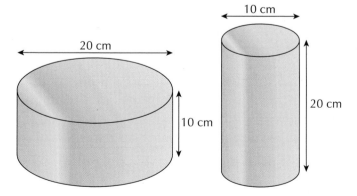

10 cm

20 cm

10 cm

20 cm

Compare the volumes of the two cans.

31 The total of Roger's age and Saleem's age is 61.

The total of Saleem's age and Teri's age is 89.

The total of Teri's age and Roger's age is 76.

Find the total of Roger's age, Saleem's age and Teri's age.

32 Mick has lasagne, a banana and a can of cola for lunch.

On a website he finds this information about the number of calories in what he has eaten.

Item	Quantity	Calories
Lasagne	1 serving/400 g	588
Banana	Medium/150 g	144
Cola	1 can/330 g	142

He estimates that he had 500 g of lasagne and a large banana with a mass of 200 g. Find the number of calories in Mick's lunch.

33 This is a hemisphere (half a sphere) with a diameter of d cm.

The volume, V cm^3, of a hemisphere with a diameter of d cm is given by the formula:

$v = \pi d^3$

Work out the diameter of a hemisphere with a volume of 500 cm^3.

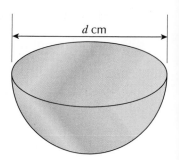

34 Three points, A, B and C, are in a straight line on level ground. There is a tree at B and a tree at C.

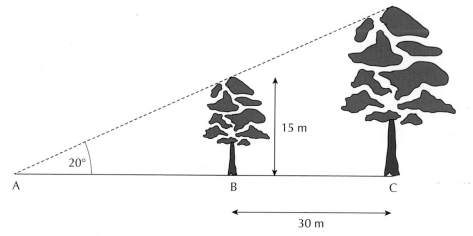

The trees are 30 m apart. The tree at B is 15 m high. The angle of elevation of both trees from A is 20°. Calculate the height of the tree at C.

35 The pie charts show the ages of men and women in a group of residential homes.

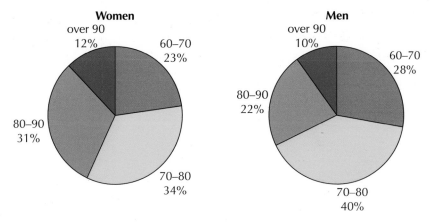

The ratio of women to men is 5 : 3.

There are 30 women in the over-90 class.

How many men are there in the over-90 class?